This book describes the structure of simulators suitable for use in the design of digital electronic systems. Such systems are increasing rapidly in size and complexity, and the use of simulation provides a means to 'prototype' a design without ever building hardware. Other advantages over hardware prototyping are that sections of a design can be simulated in isolation, and that all internal signals are available.

This book includes the compiled code and event driven algorithms for digital electronic system simulators, together with timing verification. Limitations of the structures are also discussed. An introduction to the problems of designing models is included, partly to point to how user models might be constructed for application specific integrated circuits (ASICs) and so on, and partly to expose the limitations of the modelling process.

Simulators have two functions. The first is to confirm so far as possible that a design meets its specification. The second is to check if the test program will find a sufficient percentage of possible manufacturing faults. In the former case the user must supply test vectors. In the latter, tests can be generated by automatic means. As a guide to the use of simulators the book includes chapters which introduce the subjects of testing and design for testability. A major chapter is devoted to fault simulation. Finally, the text has an introduction to hardware accelerators and modellers.

The book is suitable for electronic engineers using digital techniques, including undergraduates using design software, and postgraduates and practising engineers using simulation for the first time. It will also be useful for computer scientists needing an introduction to simulation techniques.

Electronics texts for engineers and scientists

Editors

P. L. Jones, *Electrical Engineering Laboratories, University of Manchester*

P. J. Spreadbury, *Department of Engineering, University of Cambridge*

Simulation in the design of digital electronic systems

Simulation in the design of digital electronic systems

JOHN B. GOSLING

*Formerly Senior Lecturer, Department of Computer Science,
University of Manchester*

Published by the Press Syndicate of the University of Cambridge
The Pitt Building, Trumpington Street, Cambridge CB2 1RP
40 West 20th Street, New York, NY 1001-4211, USA
10 Stamford Road, Oakleigh, Melbourne 3166, Australia

First published 1993

Printed in Great Britain at the University Press, Cambridge

A catalogue record for this book is available from the British Library

Library of Congress cataloguing in publication data

Gosling, J. B. (John B.)
    Simulation in the design of digital electronic systems/John B.
    Gosling.
        p.   cm. – (Electronic texts for engineers and scientists)
        Includes bibliographical references and index.
        ISBN 0 521 41656 6. – ISBN 0 521 42672 3 (pbk)
        1. Digital electronics.   2. Electronic circuit design–Data
processing. 3. Computer-aided design.   I. Title.   II. Series.
TK7868.D5G666   1993
621.3815–dc20   93–14882   CIP

ISBN 0 521 41656 6   hardback
ISBN 0 521 42672 3   paperback

Contents

Contents

8.5	False paths	198
8.6	Use of timing verification	202
8.7	More complex models	205

9	**Fault simulation**	207
9.1	Introduction	207
9.2	Reducing the problem size	209
9.2.1	Static reduction of tests	210
9.2.2	Dynamic reduction of the Nishida cube	211
9.3	Parallel methods of fault simulation	212
9.3.1	Single fault propagation	213
9.3.2	Extension to include faults in the wiring	214
9.3.3	Assessment	215
9.3.4	Parallel pattern single fault propagation (PPSFP)	216
9.3.5	Extension to wiring faults	218
9.3.6	Evaluation	218
9.3.7	Multiple values	219
9.3.8	Parallel fault simulation	220
9.3.9	Fault dropping	221
9.4	Concurrent fault simulation	223
9.4.1	General description	223
9.4.2	Detailed example	225
9.4.3	Change of input	228
9.5	Parallel value list (PVL)	230
9.5.1	Detailed example	231
9.5.2	Change of input	233
9.5.3	Comment	234
9.6	Assessment of simulation methods	235
9.6.1	Parallel methods of fault simulation	235
9.6.2	Concurrent fault simulation	236
9.6.3	Deductive fault simulation	237
9.7	Some alternatives to fault simulation	238
9.7.1	Critical path tracing	238
9.7.2	Statistical methods	239
9.7.3	Block orientated fault simulation	240
9.8	Timing in fault simulation	240
9.8.1	Delay faults	240
9.8.2	Oscillations and hyperactivity	241

Preface

In attempting to come to grips with the problem of designing a simulator the author found very little in the way of overall descriptions of what a simulator is, what it does or how it works. The required information can be winkled out from many different sources, but not all are easily available. This book is an attempt to bring together in one place a comprehensive introduction to all aspects of simulation in the design of digital electronic systems.

The text begins with an introduction to the purpose of simulation, types of simulation and some of the problems that are encountered in the use and design of simulators. It continues with a brief review of computer aided design suites in order to set simulation within its overall context.

In order to use a simulator it is necessary to prepare test information. To get the best out of the simulator it is necessary to adopt good design techniques. Hence the next two chapters give an introduction to design for testability and to test program generation. These are followed by a brief description of the preparation of test programs using the VHPIC high level design language (VHDL). These three chapters are just an introduction for completeness in the book as a whole, and the reader is referred to much more comprehensive texts for a proper treatment.

Chapters 6 to 9 are the meat of this work. Chapter 6 describes the two main types of straightforward simulator and gives some examples of their use. That is followed by a description of a method by which the necessary models can be written. Chapter 8 deals with timing verification and Chapter 9 with fault simulation. So far as is known this is the only introduction to modelling or to timing verification in one place. Chapter 6 is also the most comprehensive description of simulators known.

Whilst in some ways Chapters 4 and 9 go together, the topics are put in the order as here because that is the order in which the user would meet the techniques in a typical design exercise. That is to say, a designer must consider design for test and testability very early, but will then do much functional simulation before beginning the test program generation and fault simulation in earnest.

Although mentioned in Chapter 1, circuit simulation and switch level modelling are not described in detail. Circuit simulation uses very different techniques to digital work and is beyond the scope of this book. Much switch level simulation can be handled by the techniques described in Chapters 6, 8 and 9, but some additional modelling considerations become important. Some hints to these are given but a full description is omitted for brevity.

The intended readership is undergraduates in Electrical and Computer Engineering and those studying computer aided design of electronic systems in Computer Science. It is aimed at both users of simulators and at those who may wish eventually to be involved in their design. Whilst some sections of the book will be more important to one group and other parts to a different group, the whole book is relevant to all. An understanding of the working of the simulator will lead to better use of it, and an understanding of the needs of users will lead to better design. In particular, whilst users will normally use models supplied with the CAD suite, they will still have to write high level models for their own design, possibly at a fairly detailed level.

A minimum of assumptions about the knowledge of the reader has been made. The primary assumptions are that the reader understand the basic operation of a gate and a flip-flop, although Chapter 1 contains a brief description of the working of a flip-flop. It is assumed that the reader understands the concept of there being a delay between application of an input to a circuit and the output changing. An indication of how this can become more complex is given. An awareness of the concepts of set-up and hold time would be helpful but not essential. Attention is drawn to the distinction between the terms 'latch' and 'flip-flop' given as a footnote in Section 3.2.

Some understanding of simple computer data structures will be helpful, as will some elementary knowledge of electronic circuits – current flow, input and output limitations etc. In both cases readers without this knowledge will need to take some statements on trust but should not find the lack of background serious. Diagrams of logical devices more complex than a gate are very few and use the dependency notation. Readers unfamiliar with this should investigate it as it makes the function of modules very easy to understand.

Conventional drawing conventions are assumed. That is, signal flow is left to right and top to bottom unless otherwise indicated. Logic is drawn as nearly as possible to the IEEE standard on dependency notation. Digital signals are usually written as *1* or *0* as appropriate to indicate 'active' and 'inactive' respectively. Signal values are written in italic fount to distinguish from numbers or literal signal names.

Some of the references are used in the text, but others are added as pointers to additional reading. In many cases some comments are included. The selection of references is inevitably that of the author. It is limited to those which describe basic ideas well or which, in the author's opinion, are most likely to be of long term use. Since early papers often come into the former category there are rather more of them than might seem proper at first sight.

There are no tutorial questions provided as such. As Chapters 3 to 5 are introductory only, the reader should refer to specialist texts. In Chapters 6 to 9 some simple examples such as full adders or even just a two to one multiplexer can be used. It is not difficult to work out from knowledge of these devices what results the simulation and model should give and thus provide a check on the working of the example. In some places, notably in Chapter 9, some extensions of the example in the text are suggested and results provided.

Acknowledgements

The author owes a considerable debt to Miss Hilary J. Kahn of the University of Manchester, with whom he has worked both as a user and as a designer of simulators over many years. Dr Andrew F. Carpenter, also at Manchester, has commented on the initial text and has assisted with a number of matters, in particular with the VHDL and fault simulator work. Daniel Cock also made several very important contributions in relation to fault simulators. Peter L. Jones, the publisher's series editor and another former colleague in Manchester has been most helpful throughout. The University of Manchester (through Dr Carpenter) has also provided facilities to check the VHDL and to check the models by use of VHDL. Lastly, the author's wife has suffered much and long in the process of getting the text to market.

John B. Gosling
Glossop

1

An introduction to the simulation of electronic systems

1.1   Introduction

A few years ago a well known company stated that the size of silicon chip that could be designed and built would be limited by 'engineer blow-out' – what a single engineer could hold in his mind without going crazy. To overcome that limitation, techniques for 'managing complexity' have been developed. These have included methods for manipulating data in different ways. The computer can handle large quantities of data without becoming crazed and without error, leaving only the interesting and intelligent work to the engineer.

Computer aids are not limited to chip design. It is not difficult today to produce a chip which works first time according to its specification. But was the specification correct? Thus *there is no point in designing a 10 million gate chip which works perfectly to specification if the specification is wrong*. In the late 1980s, estimates varied in the region 10% to 50% that the chip would work within its intended system (Harding 1989, Hodge 1990). This was clearly unsatisfactory, so there has been increasing emphasis on the need for *system* design rather than purely chip design.

One of the problems with building hardware is that, once built, it is not easily changed. In the case of designing on silicon, change is impossible. It is estimated that the relative cost of finding faults at design time, chip-test time, printed circuit board construction time, or in the finished machine in the field is $1:10:100:1000$ (Section 3.1.1). However, (any) software running on a computer is (relatively) easily modified. Simulation is the process by which a model of the hardware is set up in software, or better still, in data structures that are 'run' by the software. The simulation can then be used to test the 'hardware' before it is ever built. When errors are found the data can be changed and further runs made until a correct design is achieved. If budget and time to market considerations allow, it is possible to try out alternative designs – sometimes called 'playing "what if" games'. Three important comments must be made before going further.

- A computer aided design system is what it says it is – an *aid*. It is *not* a substitute for the intelligence of the designer.
- Obtaining a simulation run which does not result in error reports is *not* the aim. That would be easy – just turn off the reporting. Even without such drastic measures, the simulator can be tricked, but that would leave a very bad design. The design time is reduced, but redesign will be necessary. A simulator must be used intelligently.
- Simulation does *not* generally reduce design time. It frequently increases it, possibly by several hundred percent, since design and test errors have to be corrected. The advantage gained is in reduced commissioning time and hence probably a reduced time to market. It should also lead to a more robust design having less teething troubles and requiring less maintenance, which, as already indicated, can be very expensive indeed.

It might also be noted that simulation is not cheap. For major system designs, runs of 20 days (24 hours a day) on a single user multi-million instruction per second machine have been reported (1985[1]), and as systems become more complex things are likely to get worse rather than better. This emphasises the need for careful and intelligent use of the facility.

1.2   Four aims of simulation

1.2.1   Functional correctness

A simulator is required to give an accurate prediction of the behaviour of a good system.
A simulator is required to give warning of a faulty system.

The most obvious purpose of simulation is to check that the system as designed should perform the logical function for which it is intended. This can be termed *functional* correctness (see Section 1.4.2). In order that functional correctness can be checked, two matters must be considered.

- The correct operation must be known. This requires a specification of the function which is unambiguous, and which is clearly understood by both customer and designer. It must include a

---

[1]   Personal communication.

definition of terms, which might otherwise mean different things to the two parties.

- The functionality of the system under unusual conditions of operation may be important. This can include what happens if one or more faults appear (e.g. fail-safe railway signalling) or if input data patterns are incorrectly generated. In some cases this behaviour will be part of the specification. In others not. Checking such behaviour is *very* difficult because it requires the particular situations to be foreseen.

Consider the logic shown in Fig. 1.1, where the square boxes are pieces of combinational logic. For the present purpose it is assumed that the 'loose' inputs are set in such a way that a change of the connected input signal results in a similar change in the output. The waveforms of Fig. 1.2 show the result of a check for functional correctness. Z is seen to be A & B, and is delayed relative to the edges of A and B only to demonstrate its dependence.

Fig. 1.1. Demonstration circuit.

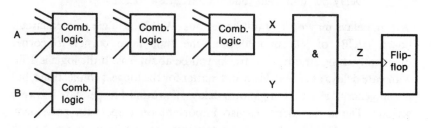

Fig. 1.2. Notional waveforms for Fig. 1.1.

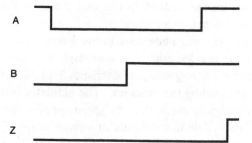

1.2.2   Speed of the system

Most systems will have some form of limit on the time they take to respond to given conditions. Even where this appears very relaxed compared with the apparent speed of the electronics it needs to be checked. For example, designs using logic with basic gate speeds of 10 ns will quite easily build up delays of 500 ns in strings of gates, and without the designer noticing. Hence it is important to check on speed of operation.

Speed of operation depends on three criteria.

- The basic circuit speed.
- How heavily the circuits are loaded and their driving capability.
- The time signals take to pass down the wires (which can be significant).

For example, a simple TTL gate is specified to have a delay of 11 ns with a load of 10 similar gates. However, it will be faster by about 0.25 ns for each gate less than 10. Hence it has

Basic gate delay   8.5 ns,
Delay per unit load (one similar gate)   0.25 ns

Wiring delays may or may not be serious. Within a single printed circuit board (PCB) of $300\,\text{mm} \times 300\,\text{mm}$, the time from corner to corner diagonally using strictly $x - y$ tracks will be about 3 ns. If the logic is TTL with gate delays as above this could matter for the longest connections, but not in general. For ECL logic with delays of around 1 ns per gate it is very serious. The wiring delay is also important on chips with polysilicon connections. The polysilicon is very resistive – say $50\,\Omega$ per square. A track $2\,\mu\text{m}$ wide and 1 mm long is 500 squares long, and hence has a resistance of $25\,\text{k}\Omega$. The capacitance is likely to be of the order of $0.15\,\text{fF}$ per μm^2, giving a capacitance for the 1 mm by $2\,\mu\text{m}$ line of 0.3 pF. The delay of a wire is of the order of one or two time constants. In this case the time constant is 7.5 ns, which is very significant.

Wire delays are unknown until a late stage in the design process. In the early stages a guess figure may be used. For later stages it is necessary to have software capable of extracting wire characteristics from the layout and feeding them to the data used by the simulator. The simulator is then run with these additional delays to check that the speed specification can be met. The feedback of layout data to the simulator is one example of what is known as **back annotation**.

Referring back to Fig. 1.1, Fig. 1.3 shows the signals X, Y and Z produced by a simulator which considers timing as well as logical operation. It is assumed that the output of a box changes several time units after its input and that all boxes have the same delay. Thus a change passing from A to X suffers about three times as much delay as one passing from B to Y. Fig. 1.3 shows that Z contains a short pulse (compare with Fig. 1.2) even though B changes later than A.

1.2.3 Hazard detection

Fig. 1.3 illustrates why timing is important. Fig. 1.1 shows Z being used as the clock input of a flip-flop. The function suggests that the pulse shown in Fig. 1.3 should not be there. However, it is, and will probably cause the flip-flop to trigger incorrectly. This is an example of a **race** or **hazard**.

Other timing constraints include ensuring that

- data to a flip-flop does not change just before the clock (set-up time),
- data to a flip-flop does not change just after the clock (hold time),
- timing to dynamic memories meets the chip specifications under conditions of tolerance (RAS, CAS, R/W etc.).

It is important that the timing specifications should not be violated during normal operation of the equipment, and so the simulator should be able to detect and report if any such circumstances occur. However, this raises an issue in relation to simulator operation, since *when a simulator finds a timing fault there is no way of knowing what the real logic would do.* The question is:

Fig. 1.3. Waveforms for Fig. 1.1 – accurate timing.

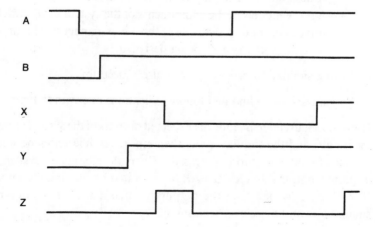

Should the simulator

- try to mirror the real logic?
- set an unknown state and/or stop?
- guess the designer's intention and try to carry on?

1.2.4 Expected outputs for test and fault simulation

Once the system can be shown to be according to the specification it will be necessary to develop a test program which will give an output that is different from the good system for as many faults as possible. This will be run on a machine to test the production hardware. To discover which faults are being tested a **fault simulation** is run. In this a fault is introduced into the good network and a simulation run. The outputs are compared with the output of the good system. The procedure is repeated for every possible fault. Simulation of every possible faulty system is an awful lot of work and special techniques are available to help (Chapter 9).

Once all the tests have been run against all faults there will be a list of untested faults. New tests are generated to check for these. Another purpose of simulation will be to determine the outputs of the good circuit for each new test as well as to fault simulate on the new tests.

It is important to notice that when these simulations are being run the system is presumed to be functionally correct. This is not true of the first three aims.

1.3   Components of a simulator

The process of simulation requires three sets of data and a program.

- A description of the system to be simulated.
- A description of the inputs to the simulated system.
- A set of models of the components of the system being simulated.
- A mechanism to process these three in a manner which mimics or 'simulates' the system being designed.

Following simulation proper there is also a need for

- assistance to find and follow indications of errors to their source.

It is often felt useful to include the expected outputs of the system along with the inputs, and the simulator will then report on differences between the outputs found and those specified. When developing a system these 'expected' outputs should *NEVER* be generated by running the simulator. If the logic is wrong then the outputs produced will be wrong and the simulation will be useless (Sections 1.2.1 to 1.2.3).

During test generation the design is regarded as functionally correct. Hence the expected outputs for the test fixture to check against can be derived by simulation. The same is true for fault simulation (Section 1.2.4).

A **model** of a component is a representation of its behaviour in a form which the simulator can use. This definition will do for now. It will be expanded in Section 1.5 and Chapter 7. The models of the components used by the simulated system may be built into the simulator in some way. However, the simulator is much more flexible if the models are held separately in a library available to the simulator. In this way new components can be added with relative ease and without having to recompile the simulator itself. The number of possible components which a simulator may at some time wish to use will run into thousands. It is better if only that subset of models actually needed by a particular network is in the computer memory at run time. This may be only 10 to 20 models, possibly less.

1.4   Levels of simulation

1.4.1   System design

As a rule a customer will present the design engineer with a specification of the system. The engineer will first divide this into functional blocks – memory, CPU, control, etc. as shown in Fig. 1.4. Each of these major blocks can then be specified clearly and passed to different people for more detailed work. Eventually the individual circuits such as 4-bit arithmetic logic units (**ALUs**) are designed. This is described as a **top-down** procedure.

However, it may well be apparent at an early stage that certain circuits or small blocks of logic (e.g. the 4-bit ALU) will be required. It is possible to do detailed work on these before their specific place in the system is fully defined. This is **bottom-up** design. Indeed, these blocks may be so commonly used as to be available as building blocks even in chip design.

In practice both high level and low level design are likely to happen at the same time. Detailed work goes on with ALUs, shift registers etc., whilst the system architects work on their inter-relationship. This has become known as 'meet-in-the-middle' design.

In order to clarify the interface specification of the major blocks of the system it is necessary to simulate these blocks without knowing their detailed internal structure. A hierarchy of models reflecting the hierarchy of the design is required. The high level blocks will be simulated with high level functions. For example, a multiplier might be simulated with a statement such as

$$a := x * y;$$

in PASCAL. Hardware description languages such as VHDL (Chapter 5) usually have provision for use of such statements. This one makes use of the multiplier of the computer rather than some simulation description. Care is required for two reasons.

- If one is designing a 32-bit multiplier and simulating it on a machine with a 32-bit number system, the results required may differ from those of the machine. A particular case occurs if the machine works with signed numbers and the new design with unsigned numbers.
- The effects of special operands. For example, $-2^{15} * -2^{16}$ is not representable on a 32-bit machine, but $-2^{15} * 2^{16}$ is (true complement number system). What does the system being designed do in such cases, and how does that compare with the results from the machine doing the simulation?

As the design progresses, simulation will take place at progressively 'lower' or more detailed levels, at least until the logic in terms of gates is reached. At this level the system is modelled in terms of ANDs, ORs and NOTs, and possibly flip-flops.

There are two issues here.

- The system has to be described separately at each level of the design, and these levels must be proved to be equivalent.

Fig. 1.4. A system hierarchy.

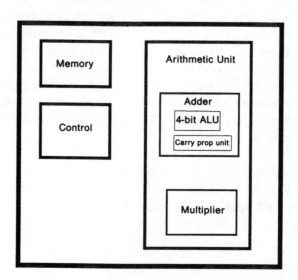

■ The design may not progress through the levels evenly, so there is a requirement to be able to simulate some blocks as high level statements and some at gate level.

The solution is to design a simulator which can handle all levels of design. It should be possible then

- to check a low level simulation against a high level one,
- to run 'multi' level simulations where different blocks are at different levels of description, which in turn:
  - o enables simulations to run faster, since only a few blocks are simulated in detail,
  - o enables larger systems to be simulated since high level blocks generally require less resource than the corresponding low level blocks.

Benkoski (1987) gives an example of a 4-bit adder that took 293 seconds to simulate at a component level (Spice), 0.5 seconds to simulate as four 1-bit adders and 0.1 *milli*seconds to simulate as a single 4-bit adder. The use of multiple VHDL architectures for one logical device is useful here, see Chapter 5.

1.4.2 High level

This is the level of description equivalent to the high level language procedure in programming. It may cover more than one level of design development. Fig. 1.4 showed a complete system containing memory etc. and an arithmetic unit (AU). This might have a functional description in which its function is described in terms of addition, subtraction, division etc. Within the AU are a series of smaller boxes. In particular the adder is shown, and this is described in terms of 4-bit ALUs, and carry propagate units. A multiplier is also shown. This is probably described in terms of 'carry save' adder units (what these are in detail does not matter here) and carry propagate units.

The words 'functional' and 'behavioural' are frequently used in this area. There is no universally agreed definition of what these mean. This book will follow Abramovici *et al.* (1990) using the following (see also Section 7.1):

Functional describes the logic function only and no timing.
Behavioural describes logic *and* timing.

1.4.3 Gate level

This level of description uses models of 'simple' gates, though in practice the models are anything but simple. The logical elements may include multiplexers and flip-flops, though both may be described in the data books in terms of ANDs and/or ORs. Fig. 1.5 shows one possible form of a 1-bit ALU. Care must be exercised in using such data book equivalents, since the real circuits frequently make use of unconventional circuit techniques. Special models not constructed of ANDs and ORs should always be considered in these cases.

1.4.4   Circuit level

Circuit level simulation is essentially different from all the other levels. The circuit is described in terms of resistances, capacitances and voltage and current sources which are the models. A set of mathematical equations relating current and voltage is set up and solved by numerical techniques. With gate and higher levels of simulation one of two real voltage levels is assumed, together with some 'unknowns,' and there are no mathematical equations to solve.

Circuit simulators typically can handle only a few hundred circuit components and nodes, since they require large in-store data structures and large amounts of computing resource. They cannot handle complete integrated circuits, never mind whole systems. Furthermore, they cannot cope with linking to higher level simulators. Usually they are used to characterise relatively small blocks (e.g. gates) which are then redescribed for the higher level simulators.

Fig. 1.5. One bit of an ALU.

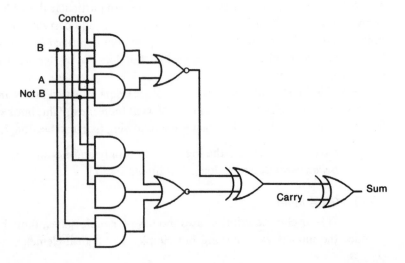

Circuit simulators give analogue results. That is, the transient response is seen as edges with real rise and fall times, and, where relevant, oscillations. Other simulators are for digital circuits only, and the results give delays, but the edges are of infinite speed (zero time). Circuit simulators can also give frequency domain results – that is they compute the zero frequency working point, substitute the small signal parameters into device models, and compute frequency responses.

1.4.5 Switch level

Between circuit and gate level there are a number of modelling methods with the trade off between speed and accuracy at different points. The primary technique is to regard the transistors as switches that are either open or closed. These techniques are known as switch level, therefore (Bryant 1984).

At its simplest, no timing is involved. The circuit is divided into sections starting from a gate output and proceeding to the following inputs. The wires marked as area *P* in Fig. 1.6 are a complete set. The switches producing signals A, B and C are included in *P*, but not those producing signals R and S. They can be simulated as a unit. They affect no other set of wires except via the gate/base of the driven transistors and no other wires affect them except via the outputs of the driving gates. Thus the size of the problem has been reduced.

Timing may be included in switch level simulation. One form of this is to estimate the capacitance of each wire, the driving gate output and the driven gate input, together with the source resistance of the driving gate(s). Algorithms for combining the time constants to obtain a delay have been developed. These may involve solving differential

Fig. 1.6. Section of circuit for switch level simulation.

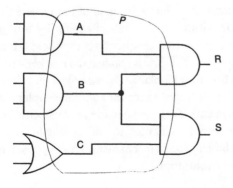

equations for the sub-networks. Mechanisms to handle feedback are also known.

Many such simulators have been described, and some are available commercially. Some give pseudo-analogue results, showing real rise and fall times of the waveforms. Sometimes they are described as circuit simulators and compared for speed and accuracy with true circuit simulators. They are much faster, of course. They can handle only the transient performance of digital circuits. They cannot give frequency responses of true analogue circuits such as amplifiers. The reader is warned to beware, and must make his/her own judgement as to the fairness of the comparisons.

Switch level details are beyond the scope of this book. Many of the techniques described are usable at higher levels also. References relevant to switch level will be included as appropriate in later chapters.

1.4.6   Mixed mode

In recent times some effort has been made to try to link circuit simulation (analogue) and digital simulation. There are at least three ways in which this has been attempted.

- Write a link between a circuit and a digital simulator. The analogue part runs for a short period of simulated time and then the digital. Suitable pseudo-analogue to digital converters are used. The digital sections also feed the analogue via appropriate conversions. There is a problem of feedback between analogue and digital, since the time scales are not really compatible. The whole thing seems unsatisfactory.

- A circuit analysis system based on events has been developed (Sakallah 1985). In the normal technique the differential equations are solved for a given time step. If the changes that result are too large, the solution is unwound and repeated with smaller time steps. In this system an estimate is made of the largest time step that can be made without the signals changing by too great an amount. An event queue (see Chapter 6) is then sent a marker at that time ahead. When that time is reached the marker will cause a further computation to take place. The analogue circuit operates in relatively small partitions to keep the solution simpler (compare the switch level system). This can be made to work with feedback, and, as will be seen in Chapter 6, the analogue and digital simulations use the same basic mechanism. This approach must be worth a lot more exploration.

- 'Behavioural analogue' (Visweswariah *et al.* 1988). In this model of the system the analogue blocks are simulated by a circuit simulator and a model developed which describes the block in terms understandable by the digital simulator. The block can then be treated as a 'digital' component by the digital simulator. This sounds obvious, but, as will be seen later, writing the 'digital' model is very difficult.

Analogue simulation is also beyond the scope of this book.

We may now define a **good simulator** from the point of view of this book as one which *performs hierarchical and mixed level simulation of digital logic, together with timing error detection, independent of technology, but recognising the distinct features of known technologies.*

1.5 . Models

Simulation is about building and exercising a model of a system (electrical or otherwise) that is being designed. Different levels at which parts of the design might be modelled have been mentioned and the different purposes of the models discussed (logic correctness, timing, etc.). These criteria also affect the required accuracy of the models.

Writing accurate models of components is a difficult and skilled activity (Chapter 7). Several companies now specialise in this type of work. It has been said that to produce a model may take as long as designing the circuit in the first place.

Consider as an example a D-type flip-flop (e.g. 74ALS74) as shown in Fig. 1.7. It has to respond to four inputs, namely preset, clear, clock and data, to give an output, Q ($\bar{Q}$ is also available). An active signal on preset (low in the particular case) sets Q to *1*, whilst an active signal on clear sets Q to *0*. The situation with both active is undefined, but often leads to both Q and $\bar{Q}$ being *1* as shown here. If either preset or clear is active, the clock and

Fig. 1.7. 74ALS74 type flip-flop.

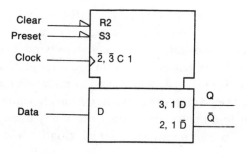

data have no effect. When preset and clear are inactive (high), an active edge on the clock line (low to high in this case) causes the signal on data to be transferred to Q. Delay from preset, clear or clock to output depends on whether Q is changing from *0* to *1* or vice versa, but not on which input is changing (the reader familiar with the gate level logic diagram might like to consider why). The model must respond as described to signals whose changes are well separated, but real circuits do not respond instantaneously, and signals must not be allowed to change too close together. The following is a list of checks to be made.

- D must not change $<t_{su}$ before the active edge of the clock.
- D must not change $<t_h$ after the active edge of the clock.
- Clock must not have an active edge $<t_{pc}$ after preset or clear go inactive.
- Preset and clear must have a minimum width.
- Clock must have a minimum period, high phase and low phase.

Few, if any, data books specify all these parameters. A further problem now arises. What should be the value of Q when any of these checks fails? What if several fail at the same time? Some solutions to these questions are suggested in Chapter 7.

An accurate model for such a flip-flop requires upwards of 300 high level programming language statements. It is to be hoped that any path through as a result of an input change will require only a fraction of these to be executed. For example, a change of the data input alone has merely to be noted – no action is required, and a clock change while the preset or clear are active can be ignored.

1.6   Test program generation

Once a simulator is in place, a network described and models are available, the network must be exercised. This is done by means of a test program. The goal of the test progam is to exercise the logic for two purposes.

- To show that the design is correct and the system performs according to specification. The test program must be agreed between designer and customer. It must cover all individual operations that the system is to perform and a range of data for each operation. It must also be tested on unexpected conditions and unexpected behaviour. For example, the reset function should set the system to a known state regardless of any other system inputs. It would be easy simply to test for initialisation with one set of data on the other pins, say all 0s. However, *if the design is faulty* the system may not initialise correctly with other data on the inputs, say all *1*s. One cannot test with all possible data sets, but a judiciously chosen set of tests should be performed.

This set of tests is used during the design phase, primarily in simulation, to demonstrate that the *design* is correct.

- A second set of tests is required to detect faults in the implementation. For example, if a wire has a break in it due to faulty manufacture, then the signal on the 'driven' end of this wire will probably be at a fixed logical level. For TTL circuits, that will generally be the higher voltage level, for instance see Fig. 4.1. Ideally one would like a test for every possible fault or set of faults in the system. Unfortunately that would be a very large set of tests. It would also be prohibitively expensive to use. Methods of limiting the number of tests will be described and assessed.

This second set of tests is needed to give a thorough test of each system built. The data sets are usually chosen to detect specific faults and are not related directly to the functionality of the system. It is not possible, therefore, to assess the functionality of the system using these tests. That was the purpose of the first set.

A further set of tests may be needed to check the system timing. Some of the tests in the first set may be provided for this purpose but further tests may still be needed. The simulator (including timing analyser here) can perform a thorough testing of the design timing. A test fixture cannot. It can merely tell whether that particular instance meets the timing criteria. Again, both types of check are needed.

Producing a test program to exercise the specified functions of the design must be done 'by hand' since only the designer understands the system. Understanding what tests to apply to check that nothing happens when nothing is supposed to happen is more difficult. Consider Table 1.1, which represents a test of the clock and data sections of the flip-flop of Fig. 1.7, preset and clear being assumed inactive. Each line in the table represents a set of inputs applied to the flip-flop and the value of the output, Q, expected after the circuit delay. It is assumed that the time between applying the inputs and testing the output is 'long' relative to the circuit delay and that the time between sets of inputs is 'long' relative to the settling time of the circuit (set-up and hold times obeyed).

The output of the flip-flop is obviously to be checked for a change when the clock changes from inactive to active (*I* to *A*, the active edge of the clock) as shown in lines 11 and 20. However, it is important also to check

Table 1.1. *Tests on a flip-flop*

Time	Data	Clock	Q
1	0	I	unknown
2	0	A	0
3	1	A	0
4	0	A	0
5	0	I	0
6	0	A	0
7	0	I	0
8	1	I	0
9	0	I	0
10	1	I	0
11	1	A	1
12	1	I	1
13	1	A	1
14	0	A	1
15	1	A	1
16	0	A	1
17	0	I	1
18	1	I	1
19	0	I	1
20	0	A	0
21	1	A	0
22	1	I	0

that the output does not change on an active clock edge if it should not, lines 6 and 13. Nor should the output change on an inactive edge under any circumstances, lines 5, 7, 12, 17 and 22. It is also important to ensure that the output does not change when D changes, regardless of the state of the clock *OR* the output. All eight possibilities can be found in this test program. In this program the first two lines are necessary for initialisation, and only four checks are duplicated in setting conditions for other necessary tests. This length of program for such a 'simple' device, and not including effects of preset and clear, or checks for possible timing faults, demonstrates the difficulty when many thousands of gates are involved.

The really big problem in test program generation is, how do we know that the set of tests is comprehensive? Consider a 32-bit adder. There are at least 2^{64} input combinations – approximately 18×10^{18}. If it was possible to simulate one combination in 1 ns (which it is not) it would still take 585 years to simulate all combinations.

It is not necessary to simulate all combinations of the inputs to check the operation fully. By careful selection of the input patterns it is possible to check all paths through a system with a much more limited set of tests. However, the problem can be made easier if the design is done with testability in mind. This implies making the nodes of the design controllable by the system inputs and observable from the outputs. The topic is known as 'design for testability' (Chapter 3). This itself is by no means easy, and the result may still require a lot of effort to ensure that the design is testable and that the set of tests provided is comprehensive.

1.7   Fault simulation

When a test program has been generated, it is required to check which faults can be detected. This is done with a **fault simulator**. Basically, a set of tests is run on a 'good' network and the results recorded. A fault is now introduced into the network, the simulation re-run, and its outputs compared with those of the good network. If a difference in the primary outputs is found, the fault is **detectable**.

The number of possible faults is very large. However, when a system is almost working, an assumption of only one fault is reasonable. One might hope in other cases that multiple faults would not occur in such a way as to mask each other relative to a 'single fault' model. It has been found in practice that, if a test program will detect all faults in which a single node is stuck at $0(s-a-0)$ or stuck at $1(s-a-1)$ it will detect most other faults as well. With MOS networks it may be necessary to test for transistors stuck open and stuck closed. The primary omission from this is the 'bridging' fault where two wires are stuck together but not at a fixed level. Again, experience suggests that most such faults will be found by the single stuck at fault model.

The number of possible faults in a large network, and hence the number of simulations to be run, is still very large. If each run were to take several days the cost would be prohibitive. Techniques have been developed to speed things up and include the following.

- Timing checks can be turned off.
- If a fault is detected at test 5, say, there is no point in running that simulation to completion – a further several thousand tests perhaps.
- By judicious choice of the first few tests, a large proportion of faults can be found in this way.
- There are a number of methods for detecting faults the effects of which cannot be separated.
- It is possible to run several test patterns in parallel and/or test for several different faults in parallel. Again, judicious choice of early tests reduces the run times.

Even with all these methods, fault simulation is a time consuming and expensive business. It is one which is ignored at one's peril.

1.8   Timing verification

Running a timing simulator, even with many test patterns, cannot guarantee that all possible timing error conditions have been checked. Consequently some workers reject such simulators altogether. Instead they perform a simulation solely to check logical correctness, which greatly simplifies the models. It often also allows the simulator to run faster. Once the logic is correct, they run a timing verifier.

The timing verifier examines the structure of the network and determines where the paths of groups of signals converge in such a way that a timing error might occur. These include narrow clock pulses (Fig. 1.3) and the usual flip-flop or memory timing problems. The procedure can be guaranteed to find all such problems. However, it may be very pessimistic due to including paths which would be improperly used if they became active, and may report many narrow pulses which are of no interest because they are not at critical points in the network or in the timing. There are techniques to make timing verifiers less pessimistic, but care has to be taken not to miss significant events as a result. To ensure this a verifier must always err on the pessimistic side.

1.9   Conclusion

The previous sections have introduced some of the important ideas, terminology and issues associated with the simulation of an electronic system. The remainder of the book will expand upon many of them. Firstly, simulation will be set in the context of electronic computer aided design (ECAD) as a whole. The process of simulation may reveal that some areas of the design are difficult to test. Thus, to reduce design times, the designer should consider at an early stage how the system is to be tested – before committing the design to the computer. The book will continue, therefore, with an introduction to issues involved in design for testability, testability measures and test generation strategies. It has been suggested that the cost of testing is 60–70% of chip costs, so it is clearly a topic which may not be ignored. The reader is warned that these are only introductions, and should consult other texts dealing with such topics more fully. The chapters here are intended as a guide to assist designers to progress to a better use of the simulator. A brief outline of methods of applying 'tests' to the logic description in a simulator is included.

The mechanisms involved in the simulator itself are then discussed, exposing some of the difficulties of making the simulator emulate the behaviour of real electronics, and providing some warnings as to the limitations of the results. This will include some of the problems of writing accurate device models. Whilst discussion of timing is an important feature of device modelling, a separate chapter is allotted to timing verification and its difficulties.

This will be followed by a discussion of fault simulation. In a sense this is tied in with test generation. However, fault simulation is not actually performed until the functionality and basic speed of the design have been confirmed. Thus it is felt appropriate to place this topic after discussion of simulation.

Finally the book will introduce additional features of simulation, in particular hardware simulators and their place in the tool set.

2

Electronic computer aided design (ECAD) systems

2.1   The design process

Designing any piece of equipment (not necessarily electronic) is not a straight path from start through to product, even when no mistakes are made. At each stage problems occur which may require previous decisions to be reviewed. For example, it may be found on simulation that a particular part of the design will not operate fast enough to meet the specification. That part of the design, and possibly others, will have to be reviewed until the criteria are met or proved to be impossible. In the latter case it will be necessary to reconsider the specification. Fig. 2.1 is a summary of the process, and the reader should refer to it from time to time whilst reading the rest of this chapter.

2.1.1   Specification

Every design begins with a customer specification, which describes the function the design is intended to perform. This may include not only the logical properties but the speed of operation, the output power driving capability, the capability of circuits that drive the design, the power supply available, perhaps limits on power dissipation and electromagnetic radiation, operation in the presence of external radiation etc. Sometimes it is difficult to satisfy all the criteria. Higher speeds will require higher power dissipation and faster clocks. The latter increases the electromagnetic radiation. Some criteria may not be important, or only marginally so. With certain pieces of logic the power dissipation may be minimal so that power limits are not important.

One important matter is to ensure that the specification is correct and covers the whole design. The problem of specifying sub-systems has already been noted (Section 1.1).

Ambiguity is a major problem in specification. A word which means one

thing to one person may mean something quite different to another. One
approach to the resolution of this problem is to use a formal specification
language such as Z or VDM (Camurati 1988). The specification of the
floating point section of the INMOS T800 (Barrett 1987), and of the
VIPER (Cohn) processor both made use of such languages. The main
advantage is that they remove the ambiguity. A second advantage is that
there are algebraic methods of proving identity of that specification with the
supposedly equivalent specification of the design at a later and more
detailed stage. Such methods of proof are, as yet, limited in their capability,
but will almost certainly become a major force in the future. Better still, it

Fig. 2.1. A design system flow chart.

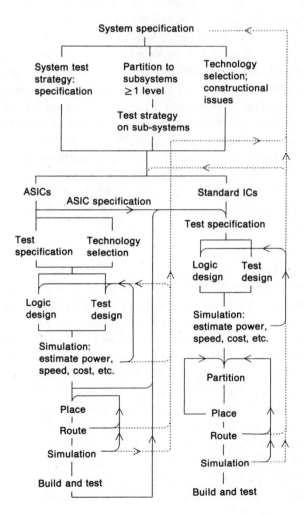

may be possible to automate the design process once the specification is in a formal language. The result should be 'correct by construction' and fast to market, even if it is not the most compact. Apart from very high volume parts, lack of compactness is unlikely to be important but time to market is usually critical. Time to market may even be the overriding consideration for high volume parts.

2.1.2   Partitioning the design

Once the system is specified it may well be that it is too complex to be designed in detail by a single person. A single rack of equipment in modern technology can contain many hundreds, or even thousands of integrated circuits (ICs), and if application specific integrated circuits (ASICs) are involved, each can contain from a few thousand to a million or so gates. All these numbers will rise as time progresses. The sizes are well beyond the capability of one person to comprehend in detail at any one time and time to market will not permit sections of the system to be designed in sequence. Hence it is necessary to split the design into smaller pieces in a hierarchy until sections small enough for an individual are formed. Each sub-design must be carefully specified in terms of its terminal functions, its speed and possibly of its physical design. The most difficult part of this process is not the hardware or software, but the interface between the *people* in the design team. In one large design known to the author, a 64-bit data highway between two sections of logic was named SH00-63 by one designer and SL63-00 for corresponding bits by another. The two designers sat at adjacent desks in the same room.

2.1.3   Test strategy

At the same time that the functional specifications are being drawn up, the test strategy should also be specified and the test specification for each block of logic produced. It is important that the test strategy is considered at this time. Adding test logic at the end of the design process is always difficult and expensive. It almost always results in unacceptable penalties in speed. Including its consideration here leads to economies in the logic and superior performance in all respects. It is still not common at present, but becomes more critical as systems become larger.

This set of tests is not the final set, since it may require modification to ensure the initial conditions for a particular test are appropriately set, or some adaptation to particular detailed design conditions. At this point decisions should be made on how to make registers and/or control signals visible outside an IC or a PCB, how to initialise the system, whether to use a self-test strategy and whether this is to be at several points in the hierarchy or just at system level, and so on.

2.1.4 Constructional issues

Also at this point in time the technology of the design should be selected. This is not always a straightforward decision. It involves factors like cost, time to market (ASICs need to be designed; off-the-shelf ICs do not), whether it is important to use ASICs for privacy of the design, small size, weight or lower power, type of packaging of ASICs or standard ICs (small outline, etc.), racking, cooling, operator requirements and, by no means least, aesthetic appearance.

Finally, arrangements for operator control and maintenance must be considered. One always hopes equipment will operate correctly for ever. Real life is rarely so pleasant. The design must be such that faults can be located rapidly and corrected, even if 'corrected' means exchanging PCBs. The cost of repair in the field is very high; three orders of magnitude higher than the cost of getting things right at an early stage in the design. The effort at this point in time gives a very high pay back later, though this may never appear on the company balance sheet.

2.1.5 Logical design

The next step in the design process is to perform the logical design. For those approaching a design for the first time this is often the starting point. For example, logical design exercises in the early part of an educational course usually require the detailed design of some piece of logic or a circuit. It is vital that the reader realises just how much time and effort has to go into the overall system design before this point is reached (see Fig. 2.1).

Precisely what is involved in logical design will depend on earlier decisions. It will almost certainly require putting together a number of integrated circuits. Today it is rare for this to include more than a very small number of simple gates (NAND, OR, INVERT, etc., but see below). It is more likely to include microprocessors, memory, shift registers, bus drivers, arithmetic logic units (ALUs), and other major blocks, together with devices for controlling the system. These will include programmable logic devices (PLDs).

Again, it is worth sounding a warning. It is quite normal to begin a design with the data paths, and feel one has almost finished very quickly. In terms of the percentage of the logic involved one is well on the way. However, the last 10–20% of the logic will require 80–90% of the design effort, for two reasons.

- The *sequencing* of the functions of the design and general control. This usually requires examination of the fine grain timing of the logic (memories, flip-flops, etc.) and ensuring that the many restrictions are met under both fastest and slowest conditions of the control logic. Indeed, one may have to consider the fastest path through some logic and the slowest path through other parts. This is an area in which it is very difficult to spot all the critical conditions, and is thus an area where particular care should be taken during simulation (Chapter 8).

- *End effects.* An example is what to do with an arithmetic overflow. The dangers are illustrated by the following anecdote. It is said that a prototype computer was used in the design of a rather unusual roof for a building. Just before it was to be built an observant civil engineer noticed something wrong with the results of the calculations. The problem was eventually traced to the fact that the arithmetic overflow logic of the computer had not been connected – and obviously not tested. The calculations had just ploughed on with a large number wrapping round to become a small one. Without the observant engineer *this would have cost lives.*

Thus there is a need to take care over such matters, but, as with late consideration of testing, failure to consider them at an early stage can be costly in logic and may also cause the design to run more slowly. In another example, a design for a multiplier involving several clock cycles was laid out very carefully to ensure that it would run as fast as possible with minimum wiring effects. However, there was a small piece of logic tidying some end effects of each cycle which was placed 'in any free space.' The delay down the slightly longer wires to this piece of logic caused the whole multiplier to run more slowly than the main arithmetic loop, and as a result was the limiting factor on the speed of this unit.

2.1.6   ASIC design

If the system is to incorporate ASICs then design with simple logic gates may be necessary, though today most vendors have libraries of higher level functions including basic microprocessors often referred to as **cores**. On the other hand, not all functions for every design will exist. The designer will certainly work with gates, and may work with transistors to produce basic functions. The latter may well be designed at switch level, and will not require detailed knowledge of circuit design techniques. Each ASIC will need to be designed as a small 'system' on its own using the principles described earlier. The full system design should have resulted in a specification appropriate to the chip environment. Provided this is met the design should be suitable. Matters of speed, end effects, control and, above all, testability must be considered. The reason for the latter is that methods of probing the internal circuit of an IC for testing purposes are extremely expensive to provide and operate. Few design teams can afford them and normally they should be assumed to be unavailable. Even when they are available they are still a last resort.

The logic design of PCBs is not essentially different from that of designing ICs. Each IC on the board is comparable to an ASIC macro on the silicon. Indeed, the logic in an ASIC today is equivalent to that which would have been put on a PCB using standard ICs yesterday.

2.1.7 Interfaces and pin limitation

As a matter of experience, however many pins there are available on a chip or PCB, some designer will want more. Except where a complete (sub)system fits on a single chip or PCB, the number of pins rises in proportion to the number of gates to a power between 0.57 and 0.75 (Landman and Russo 1971). In particular, data highways require large numbers of pins. For example, consider a 32-bit processor. There is a requirement for a 32-bit bidirectional connection to the processor. However, there is also a need to be able to specify memory addresses. Today these, too, require 32 bits. There will then be a requirement for a selection of control signals, interrupt lines and priority signals etc. The total easily reaches to over 100. If floating point facilities capable of handling IEEE double format are to be included, then 64-bit numbers are essential. To supply these as two 32-bit ones is slow. Processors with 64-bit integer and address highways are now appearing.

In assessing pin requirements it is easy to forget the requirements of power, or to suggest this is just two pins. There are two reasons why this is a mistake.

- Power. A modern PCB may well consume several amps of current, or even several tens of amps. Most PCB pins are limited to one or two amps for proper performance. If inductive and resistive drops, heat in the pins and metal migration are to be restricted, then an adequate number of pins is needed for both the power and the ground supplies. The numbers may be different, but similar restrictions apply to ICs. A 10 W IC with a 5 V supply requires 2 A, and IC pins cannot handle this much.
- Noise. Consider a 32-bit data highway at a time when all signals

are changing together – which does happen. The ground return has to handle the current of all 32 bits. TTL bus drivers typically drive 16 mA, which multiplied by 32 gives over 0.5 A. This has to pass through the ground return. If the pin inductance is 5 nH and the current changes in 15 ns then the noise voltage on the ground line is 170 mV, which is large enough to cause a spike on a clock line. The resistive effect will generally be much less.

Providing several ground pins reduces this. For ICs the currents are smaller but several ground pins should be provided. Very often in ICs the pins supplying power to the pin drivers are separated from those supplying the internal logic. Thus large current changes due to output driver switching do not affect the more sensitive internal logic.

As with other matters, these must be considered at an early stage in the design, since, if there are too many pins for a sensible package, then some revised partitioning of the design is necessary.

2.2   Design capture

As the design progresses it will be captured in a computer data base. There is a school of thought which prefers a textual description as more precise than diagrams. Standard hardware description languages such as VHDL (Chapter 5) are textual. Whilst the data internal to a computer may well be in that form, most engineers prefer a picture. Indeed there are systems available where the input can be in graphical form. Whether captured in graphical form or not, it is usual to produce graphical representations.

In most cases the CAD software will contain libraries of components. In the case of ASICs this will be predefined blocks of logic, but the principle is the same. Each component will have several representations in the system.

- A graphical representation.
- A pin list.
- A model or models for simulation. Multiple models may be used for simulating at different levels of abstraction.
- Physical data – where to place pads on a PCB for a commercial component, or IC mask details for an ASIC macro.
- Electrical data to allow checking of compatibility with other signals connected to its 'pins.' For example, high and low logic levels and current source and sink capability of outputs.
- Power/current consumption figures, etc.

This data is used in the different parts of the CAD suite.

The design capture software may have several outputs.

- A **net list**. This is a list of all pins (or pads for an IC macro) connected in a given net. A **net** is simply a number of places in the design that are connected together. The net list may or may not be available in a textual form. For PCBs a print-out is almost essential for fault finding.
- Logic diagrams, with cross references where a signal is used on several different sheets.
- Parts lists, probably with ordering information.
- Total design power and current requirements, and separate figures for each ASIC and PCB.

An important consideration is how the CAD system knows what is connected to what. As the designer enters the data each signal is given a name. In a textual description this should be something meaningful. For example, if a division by zero is to be detected, the signal indicating this might be named DIV_BY_ZERO. A name such as '***????' (an actual example) is not helpful! If a graphical input is used then signals may be named in a random fashion by the software. So long as the designer does not need to see them this is acceptable.

When the designer joins two IC pins (or equivalent in an ASIC) the system assigns the same name at both ends of the connection. In the textual form this is done by the designer giving the two points the same name. Thus, by scanning the list of signals connected to all pins, a list of pins by network connection can be built. For example, Fig. 2.2 shows four ICs, which may be on one drawing as here, or on separate drawings, or could be listed as (for example)

IC1 ALU [a0, . . a3,b0, . . b3,sum0, . . sum3,G3,P3]
IC3 CGEN [G3, P3, G7, P7, ... C7]

etc.

The system library knows the IC type 'ALU' and contains a list of signals in a known order, together with the pin numbers. In this case the first signal could be pin 1. By positional association this will be signal a0. The next pin in the library list may be 3, associated with a1. Other associations follow similarly. It is not usually necessary to specify the power and ground pins in the list above. Fig. 2.2 shows a limited selection of pin designations for clarity.

The details of all ICs are provided by the designer, enhanced from the library and placed into a design file. A net list can be produced. At some point the software picks up the signal sum_0 and searches for all occurrences of it. So far as this diagram is concerned it finds two places, an output and an input. Thus a net list of the following form can be produced.

sum_0   IC1 : 12, IC4 : 3, ...
G3   IC1 : 16, IC3 : 1, ..
P3   IC3 : 2, IC1 : 17, ...

etc.

The same piece of software will probably perform some **electrical rule checks** (ERCs) such as ensuring that a network contains only one signal source OR all signal sources are capable of being connected together – all TTL open collector or all tri-state devices.

2.3   Simulation

Once the design is under way and partitioned it will need to be simulated. This implies that a model of the design is created in the computer, together with a model of the environment in which it is to be operated. The environment model is usually called a test program. In its simplest form this is just a set of inputs such as might be applied to the finished equipment in the field. The system model is 'run' and selected signals displayed. It may be possible for the computed outputs to be compared to 'expected' outputs. Very often it will be possible to simulate blocks of the logic on their own for ease of finding problems once they have been isolated to that block.

Fig. 2.2. Generation of a net list.

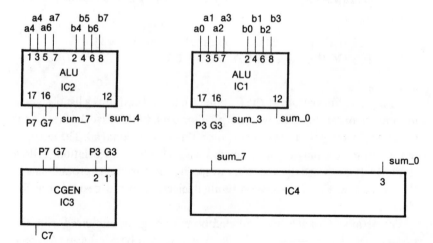

It cannot be stated too strongly that simulation will not result in the design being completed faster. It will be slower, probably much slower. However, it should lead to a design which is right first time and hence will save time and money in correcting mistakes later. It should also be more robust. It is repeated that the cost of correcting mistakes at this stage is very low compared with having to do it later.

There are many dangers in simulation.

- The check on the design is limited in extent due to pressure to finish the design work.
- The results of a simulation run are taken as the 'good results.' If logic or test program or both are faulty this makes simulation useless – or worse. That may seem obvious, but under pressure to get the design finished the temptation is strong.
- The logic is 'adjusted' to make fault reports from the simulator 'go away.' One designer working with 10 ns logic inserted a series of 1 ns delays to eliminate flip-flop timing complaints. In the real circuit such accuracy was impossible to achieve because of the wiring delays in the physical construction. A redesign was necessary. Unfortunately this was not discovered until the hardware was being commissioned – by someone else.
- Monitoring of results may be turned off. This is usually accidental, but has been known to be deliberate. In one simulator the author came across this was the default condition.
- Only sub-systems are simulated and the whole is not. Again, pressure to get the design finished may lead to the temptation to miss out the whole system simulation.
- During layout, adjustments to connections may be made. Errors in typing or otherwise in doing this will alter the function of the system. A post-layout simulation is *essential* (see Fig. 2.1).

2.4   Test program generation

Although this is being described after simulation, it will be clear that the simulation will only be as good as the manner in which the system model is exercised. It is no good just training an airline pilot to fly an aircraft on a simulator. He must also be given realistic practice at handling emergencies. It is said that the pilot of the aircraft involved in a crash in 1990 did not have such experience. The same must be true in electronics. Indeed, in some applications the consequences of failing to do so could be as disastrous as in the case of the pilot.

There are several aspects to test program generation.

- Check that the logical function of the design meets the specification.
- Check that the speed specification is met.

These are the obvious ones, and will be the designer's initial concern.

In production testing the design is assumed to be able to meet the specification provided it is correctly manufactured, i.e. it is assumed to be 'good' under some criteria. If the machine testing the finished product is to find faults introduced in manufacture, the result of testing a faulty system must be different from that of testing a fault free one, whatever the fault. This leads to a consideration of the types of fault that can occur – broken wires, wires shorted to power or ground or to each other, broken transistors – and the effects these have on the network. For example, if the input to a TTL circuit becomes disconnected it behaves logically as if the signal is stuck at the higher logic level. Thus there is a class of fault in which a circuit node is stuck at a logic value. The test program must be able to check most, if not all these faults. Hence there are further aspects to the test generation process.

- What are the faults that can occur?
- How can those faults be modelled in the simulator?
- Assess the extent to which the test program can detect these faults.
- Generate tests to find the faults not detected by the test set so far developed.

It should be realised that, when considering production tests of the system, a design exists which is defined as free of faults. A simulation of this fault free network can now be taken as the 'expected result,' and simulating a network with a fault deliberately introduced should give a result which is different to that found for the fault free network. Noting a difference can be taken as an indication that the test can find that fault.

The total number of possible faults in a network is very large indeed, and the cost of running such tests correspondingly high. If there are N nodes in the design and a fault is defined as a node stuck at 0 or stuck at 1 then the number of possible faults is proportional to N^3. For example, for a circuit of 1000 nodes, the order of 10^9 tests would be needed. At 1 ms per test (which is a long way from being possible) each system produced would take 11 days to test (24 hours per day). Clearly some compromise between cost and comprehensiveness of testing must be reached. A 1000 node circuit is *very* small by today's standards. This does not take into account other types of fault which may occur, such as two nodes stuck together.

It is also found quite frequently that certain parts of networks are difficult to test – even impossible. This leads on to the concept of designing a system to be testable from the start. In recent years, a number of techniques for doing this have been developed. Furthermore, there are procedures by which the designer may assess the ease or otherwise with which the design may be tested. Applying testability criteria to a network at any early stage in the design will ease the test program generation problem later – and reduce the time to market. Use of testability measures will enable redesign to make the network more testable, and again, if this is done relatively early in the design process, the cost in time, effort and wasted simulation resources will be reduced. Of course, these matters do make the initial design time *longer*. As stated before, attention to detail at this early stage is much cheaper than attempting to correct matters later.

One possible technique for testing which does not rely on a tester is self-testing. The 'tests' which are run need to be just as comprehensive as if they were run on a test rig, and the techniques for generating tests may be similar. There is one additional problem – the self-test logic must be able to test its own operation!

2.5 Placement and routing

Once the logic has been designed and there is reasonable confidence that it is correct as a result of simulation, the individual devices must be connected either on an IC or a PCB or both. Where sub-modules have been individually simulated, preliminary module layout can be undertaken, even though other modules may not be so far advanced.

For a system design the first step is to partition the logic between PCBs (assuming more than one is needed). This can often be done early on in the design and, indeed, needs to be so done in order that sections of logic can be reasonably self-contained. There are two possible exceptions.

- There may be a question as to which of two boards a particular interface should be placed on.
- It may be decided at a relatively late stage to place two or more relatively small modules on the same PCB.

Further partitioning may then take place to areas of the boards, and decisions made as to whether some part of the system should be on custom ICs. There are several reasons for using ASICs.

- Savings in space, power, cooling, racking, electromagnetic compatibility etc.

- To make the design more difficult to copy.
- For cheapness in high volume applications.

High volume in this context can mean just a few hundred or even a few tens of finished systems, since placing logic in ICs also results in savings of space, power, etc.

Partitioning of logic within an IC follows roughly the same procedure, where the 'system' is now bounded by the pins of the IC. However, layout of ICs has other constraints. Consider an IC which is designed so that the outputs will feed appropriate inputs of another IC of the same or different type. Fig. 2.3 shows two possible pin configurations.

In Fig. 2.3(a) the IC pins have been assigned for convenience of IC design with no consideration for system requirements. Even in this simple example the wiring space is obviously significant and there are at least two places where two wires cross. In Fig. 2.3(b) the internal connections of the IC are probably a bit more difficult, but once made are reliable. The external wiring is very simple and without doubt, more ICs can be packed into a given area. This applies to ASIC modules as well. Hence it can only be *preliminary* layout when system design is not complete.

The wiring of the individual components or logic elements is now done. If the placement of the components was good the wiring will not be too dense. However, there are a number of ways in which the layout might be improved.

Fig. 2.3. Bad (a) and good (b) pin allocation in IC design.

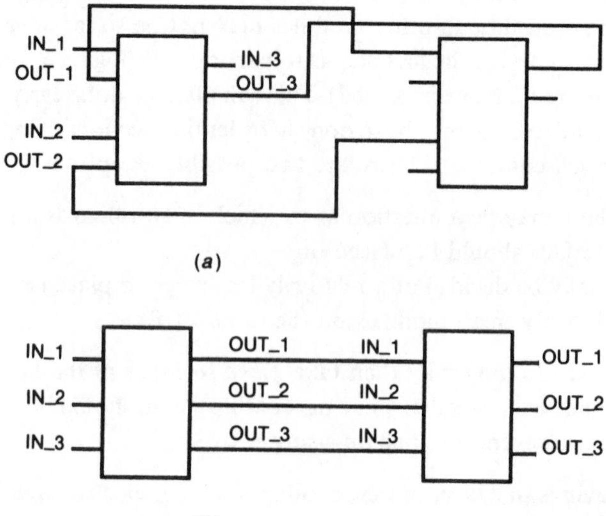

(a)

(b)

Consider Fig. 2.4(*a*). The two inputs of the gate have identical functions, so exchanging inputs will produce a better layout as shown in (*b*). In the former case, two wiring 'channels' are needed; only one in the latter. In Fig. 2.5 a similar situation applies but here two gates must be exchanged. At present, automatic software is just becoming available for this.

Placement and routing are a pair of processes which need to interact with each other. Unfortunately there is no way of confirming a good placement without doing a great deal of work on the routing, and even minor changes to the placement can require a total restart to the routing. A great deal of research on the interface between the two remains to be done. The author recommends that every student spend time laying out one substantial piece of logic 'by hand' at an early point in his/her career. The exercise will emphasise, as no amount of words can, the problems of routing and the interaction with placement that will enable further designs using automatic tools to be done with a much greater change of success, since in most cases the tools accept (or require) suitable initialisation. Or, in other words, the human brain is still superior for solving problems concerned with spatial relationships.

Fig. 2.4. Interchanging gate inputs.

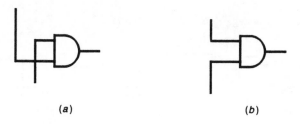

(*a*)   (*b*)

Fig. 2.5. Interchanging gates.

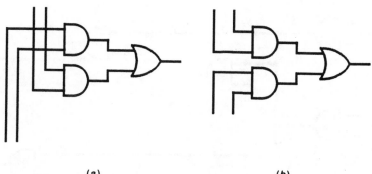

(*a*)   (*b*)

2.6   Wiring delays

Once the board or IC has been routed the length of the connections can be calculated. The characteristics of the tracks should be known to the CAD system, so that wiring delays can be calculated. With connections made in metal this will only be important for high speed logic. In general the connections can be treated as L–C transmission lines and delays are easily calculated (Gosling 1985). For connections in silicon or polysilicon, an R–C transmission line is used. The reader should attempt to calculate the equivalent time constant of a two-stage R–C filter as shown in Fig. 2.6. It is immensely difficult. Various approximations have been proposed (Gaiotti 1989). However, by far the most important factor is the equivalent circuit of the driving circuit, V_s and R_s, a fact which is not always made obvious in the literature.

Having computed these wiring delays, the values should be inserted automatically into the network description. This is known as back annotation. The system must then be resimulated with critical timing tests to ensure that the speed specification is still being met.

The number of elements to be simulated has now risen significantly. Consider the network fragment in Fig. 2.7. The layout of the diagram is intended to convey something of the physical layout. One or other of the

Fig. 2.6. Two-stage R–C transmission line.

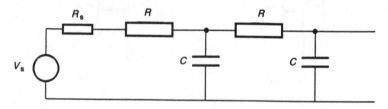

Fig. 2.7. Wiring delay network.

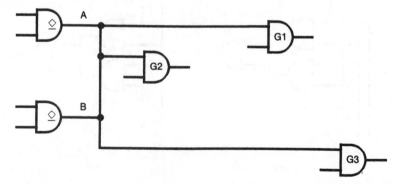

two open collector gates drives three load gates. The delay from A to G3 is different to that from B to G3. The delay from A to G1 is different again. There are at least six different delays needed here. These must be inserted as additional pseudo-elements as in Fig. 2.8, and pseudo-gates to combine them to three real gate inputs are needed. The pseudo-gates are shown as AND gates, which is appropriate to TTL logic. The network is now much larger.

For this purpose it can be presumed that the logical function of the system is correct, and only the timing is to be checked. In particular, short pulses could be generated as a result of the different delays. Hence a **timing verifier** might be more appropriate (Section 1.8 and Chapter 8).

It should be stated that the use of a timing verifier at an earlier point in the design cycle is useful. After all, there is no point in proceeding to layout if the timing of the logic without wire delays is not satisfactory.

2.7 Silicon compilation

In the writing of computer software progress was made from writing actual machine instructions to assembly language to high level languages. This might be compared to progress from designing electronic systems with gates to macros (e.g. ALUs, counters) and to processors. In software, a compiler takes a high level description and reduces it to machine code to run in the machine. By comparison, a silicon compiler takes a description of a system and reduces it to ICs, gates or transistors as required. Since the reduction is automatic it should be **correct by construction**, and the need for functional and behavioural simulators is removed. Timing analysers are still required. The automatic software should also do placement and layout. Timing analysis after layout is also still needed.

Fig. 2.8. Pseudo-network of Fig. 2.7.

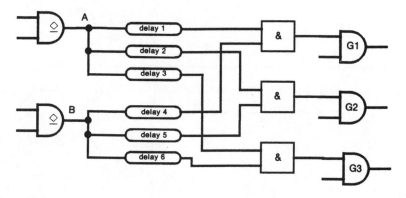

This is the ideal. The practice has not yet been fully realised, though it is being approached ever more closely. Hollingworth (1991) describes one approach. It is unable to make all design decisions, so a measure of interactive intervention is used. This gives opportunity for error, so some functional simulation is still required.

Silicon compilation makes use of formal specification a real possibility. Means to 'prove' a formal specification are available, though probably not yet useable on any but quite small systems. The VIPER chip had only 25K gates. However, they would probably work on a functional level description. Proof of that, followed by *correct by construction* implementation, should be both fast and accurate. At this point it has to be asked how the silicon compiler software itself can be proved correct! Formal specification methods cannot be used to verify timing specifications.

Several formal specification systems currently proposed do not fit the model described in the previous paragraph. To all intents and purposes they are high level design languages. 'Proof' of the design is done by simulation, usually specially written and presented as 'something different' from 'conventional' simulators. The reader is again warned to be discriminating.

2.8   Conclusion

This book is primarily about simulation. This chapter has attempted to set simulation within the context of the total design. It has demonstrated the close interaction of the different steps in the design process, especially in relation to testability and testing with simulation, and the need for simulation at several stages in the design sequence.

3

Design for testability

3.1 Cost of testing

3.1.1 Advantages and penalties

Most electronic system designers will never need to design a simulator. They will merely need to use one. An understanding of how the simulator works will enable it to be used more effectively, and avoids investing it with powers that it does not possess. However, most designers will have to write test sequences which the simulator will use to exercise the logic. They will also need to write programs for the equipment test rigs for exercising the real logic. These two activities overlap to some extent. However, checking that the system performs its specified functions is a design phase procedure and is used primarily in the simulator. Once the design is accepted as adequate it is necessary to check that any possible manufacturing fault can be detected during testing. The latter set of tests does not need to be 'understandable' in terms of the normal operation of the system since testing during manufacture is mainly on a go/no go basis. Developing and assessing the value of these tests is a major task and requires much further simulation. It is for this reason that the main chapters of this book begin with a look at the problems of writing test sequences.

The importance of careful testing of a design is illustrated by the costs involved. For the sake of example, let the cost of simulation be 'one' in whatever unit is appropriate.

Suppose now that the design is of a silicon chip, and an error is found during testing. The chip has to be redesigned. Conventional wisdom, quoted by a number of authors, suggests that the cost of this redesign will be 10 in the system of units mentioned above.

Suppose again that the fault is not discovered in the chip, perhaps due to inadequate testing. Chips are built into printed circuit boards (PCBs) and the fault discovered there. The cost of correction has now risen to 100. If the design gets into production and the fault is found by a user in the field, the cost is 1000. Worse still, a fault in the field could, in some cases, result in an accident costing lives!

Clearly there is a great advantage in spending time and money on simulation and the generation of good test programs. In an academic establishment the pressures are different from those of industry, since in a teaching exercise there is no real customer. Even in the industrial design laboratory, the pressure to get the design finished on time may tempt the designer to skimp on testing and simulation. These pressures must be resisted. One particular example can be quoted.

In the late 1960s, when small scale integrated circuits were available, the author was involved in the design of a unit consisting of about 1000 such circuits. It was built on six very large PCBs. The simulator was very slow, and, due to limited computer power, could only be run at nights and weekends – there was no such thing as interactive computing! There was much pressure to get the design completed quickly, but simulation continued to indicate problems. In spite of the extended schedule, the simulation was run to completion. The payback came when the final design was commissioned by an inexperienced research student in three days. In the same machine, another unit on which this author was involved was simulated less thoroughly, and in some places not at all. It took many months to commission. The design phase of the first mentioned section took longer than originally estimated but in the end the total time to get the unit working was reduced.

It has been suggested (Wunderlich 1987) that the cost of testing is 60–70% of the cost of designing an IC. With test synthesis techniques, the statement will still be true, but the cost may be hidden within the synthesis software. Such expense must be justification for doing everything possible *from the outset* to improve the testability of a design.

3.1.2   Problem size

To test a design for all possible faults is a massive problem. There is a divergence here between checking a design and possibly testing a prototype on the one hand and production testing a 'proven' design on the other. In the former case the designer wishes to know not only that there is a problem but also what that problem is so that the design can be corrected. Most, if not all of this type of 'testing' should be done by simulation. At this stage one is interested only in the functionality and timing of the design. The sort of errors that can occur include the following.

- Missing some data dependency.
- Ignoring a special case in the data – for example, an arithmetic operation with a result too large to be represented.
- An error in typing data into the computer. For example, typing the inputs of a block of logic as block_in_0, block_in_0, block_in_2 ... when the second input should (obviously) have been block_in_1.
- Timing errors of many types.

With the possible exception of timing errors, these faults are in no way related to the technology and, in particular, should not be confused with stuck at faults (see below).

Once the design has been proved to be functionally correct, then it is necessary to consider how to test the production units. In this case it is only necessary to know whether the unit is good or not. It is necessary now to consider the effect of the technology. Faults are introduced into the unit by the manufacturing process. What are these faults and, more importantly, how do they affect the way the design behaves? Knowing this, it is possible to design tests to detect the fault. For example, a broken track, whether on an IC or a PCB results in an open circuit at a gate input. For TTL this is equivalent to that input being stuck at the higher logic level. With ECL it is equivalent to being stuck at the lower logic level.

Tests for system functionality need to be understandable to the designer. In general they will be intelligently generated sets of data. They may well be specified in the contract with the customer, though most designers will wish to go some way beyond contractual obligations. Considerations of cost and legal liability will dictate how far, especially in safety critical applications.

Test programs for production testing will also be run on the simulator. These need not be related in an obvious manner to the design function. They must ensure, so far as is possible, that no manufacturing fault can cause faulty operation. Means to assess the comprehensiveness of these tests are required (Chapter 9).

The literature on testing ignores the need for tests which are understandable to the designer and hence can be used to find faults in the design. This author believes the distinction between these and the more extensive tests is critical to the design phase of a project. It may be that the distinction is rarely made because the former case is not susceptible to automatic analysis, generally speaking, whereas the latter is. Thus authors concentrate on the latter problem, the former being presumed 'easily solvable,' which it is not. Both types of testing are of critical importance to the success of any design.

For production testing, much of the literature is based on the premise that it should be possible to test for any fault, even down to faults within a particular gate. There must be a question over the necessity for this. We need to know the following.

- Is there any fault in the system that can cause it to operate incorrectly? If a fault does not cause incorrect operation then there is some redundant logic. This logic may be very necessary to prevent spikes happening if signals change in the wrong order. Faults in such logic cannot be detected by *static* tests, that is, tests in which signals are allowed to become stable before checking their values. Dynamic testing may also be very difficult or require very expensive testing equipment. Other than in such cases, it should be assumed that there is no redundant logic in a good design.
- Where is the fault, given that its effect has been detected? It is not always necessary to know in great detail. If a circuit fault results in the output of a gate being always *1*, then a gate output stuck at *1* fault is adequate. It may even be possible to extend the same argument to larger blocks of logic, especially when attempting to repair a system in the field.

In both types of testing, the larger the block of logic used as the minimum block the fewer and simpler the tests needed. For example, if the basic unit of a 32-bit adder is a 2-bit adder block, tests will look for faults at input and output of these blocks rather than the inputs and output of every individual gate. The balance must be carefully chosen. Use of larger blocks could be dangerous, as will be discussed later. Too many tests is very expensive in the use of test equipment. Too few tests increases redesign or field repair costs.

Reduction in tests is inevitable because a number of faults cannot be distinguished. Consider the four-input AND gate shown in Fig. 3.1. No set of tests can distinguish between the output stuck at *0* and one or more of the

Fig. 3.1. Equivalent stuck-at faults on an AND gate: (*a*) output s-a-0; (*b*) one input s-a-0; and (*c*) multiple faults.

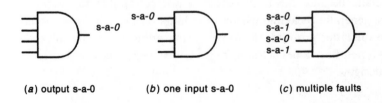

(*a*) output s-a-0   (*b*) one input s-a-0   (*c*) multiple faults

inputs stuck at 0. The effect of all such faults is identical. Thus these faults are equivalent. Fig. 3.1 shows three. The first is the output stuck at 0. In the second, one input is stuck at 0. The third has multiple faults, including two inputs stuck at 1. This approach is known as fault collapsing and will be discussed in more detail in Chapter 9.

One very commonly used method of reducing the number of tests is to assume that the design is nearly right and there will only be one fault. This means that it is not necessary to test for multiple faults. The assumption may well be valid when testing a proven design that has become faulty, but is questionable for

- prototypes
- testing chips off a production line where faults on the silicon, dirt, scratches, dislocations etc. may well cause multiple faults. In these cases it is hoped that multiple faults will still show up, and not be such as to mask all possible single faults. Abramovici (1980) and Agarwal and Fung (1981) discuss the justification or otherwise of this assumption.

Another common assumption is that all faults are of a node **stuck at 1** or **stuck at 0**, which will be written **s-a-1** and **s-a-0** in what follows. Where either is meant, **s-a** will be used. In fact, due to bad metallisation or PCB production, two wires stuck together but not at a specific value is quite possible. These are known as **bridging faults**. They may well *not* be detected by tests for s-a faults. A careful selection of test data, keeping in mind the physical layout of the design, will help here. However, the problem of bridging faults gets relatively little attention in the literature, probably because of its intractability – see Chapter 9 and references.

Another type of fault results from the extremely high input impedance of MOS transistors. If a driver never turns on then the driven signal will remain at the last set value. As will be discussed more fully in Section 4.6, this changes a combinational circuit into a sequential one, and two test patterns are required to check the operation. This type of fault is known as **stuck open**. Stuck closed faults, by contrast, almost always result in a s-a fault.

A final method of reducing the problem of large numbers of faults is to split the design into smaller blocks, each separately testable. Suppose there are 10 blocks each of 1000 nodes. The number of possible s-a faults is 10^{12}. However, if the blocks are separately testable then only $10^{\bullet}$ 10^9 tests are needed. Further, it may be possible to test several blocks at the same time (in parallel), thus reducing still further the time to test (though not the number of tests).

3.1.3   Combinational and sequential logic

Testing of sequential logic has proved very difficult. The primary reason is that it is necessary to supply several sets (known as **vectors**) of test data per test. For example, to test a register it is necessary to supply the data first, and a following vector with the clock change. This ensures that the set-up times are obeyed (see Table 1.1 for example). The general case is very difficult, so automatic test generation is virtually impossible, though specific structures are testable in a reasonable manner.

Combinational logic does not suffer this disadvantage and automatic test generation methods are known. One way to solve the sequential logic test problem is to design the system as blocks of combinational logic separated by registers. The registers are specially designed to be testable *and* to act as input to and/or output from a combinational block, giving the advantage of reduced logic size to be tested. This topic is discussed further in Section 3.3.

A form of sequential logic occurs when there are feedback loops within logic which is otherwise combinational. The problem arises because the fed back signals effectively provide second or possibly further input vectors for each vector of the primary inputs. This makes the circuit behaviour difficult to predict, especially under fault conditions.

It is questionable whether such feedback is ever good design, since it is possible to generate pulses of a width which is always subject to manufacturing tolerances. It is almost certain that a feedback system which includes a clocked device will be more reliable than one which does not. However, if such a feedback system is essential then the feedback loops need to be broken during testing. Fig. 3.2 shows a block of combinational logic with feedback loops each containing an AND gate. During testing the

Fig. 3.2. Breaking of feedback loops during testing.

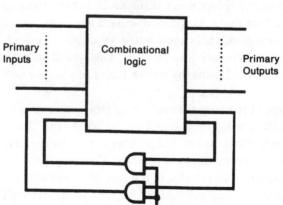

Primary Inputs   Combinational logic   Primary Outputs

0 during testing

control on the AND gate is 0, ensuring that the feedback loops are broken. If it is required to set the combinational logic inputs that are fed from here during testing, then the AND gates must be replaced by multiplexers with the other input under external control.

For synchronous designs, techniques such as scan and BILBO (see below) can be used. The design procedure ensures that the combinational logic blocks have no local feedback.

Of course, one cannot guarantee that a fault will not result in the production of feedback loops. This will happen only with bridging faults, not with s-a faults. This is another reason why testing for bridging faults is so intractable.

3.1.4 Design for testability

It will now be appreciated that generating tests is by no means simple. It does not need a great stretch of the imagination to realise that some designs, and some design styles will be easier to test than others. Hence the problem of test generation can be considerably eased if it is considered at every stage in the design process. In particular, it helps if certain additional logic is provided in some places, and if some particularly awkward structures can be avoided. The remainder of this chapter will introduce some of the methods available to the designer to make it easier to generate a test program. It is included here because it leads to a reduction in the number of tests needed, and thus to less simulation and faster design. No attempt to provide a complete guide for testability is made. The reader should consult specialist texts on the subject for that. The next chapter will introduce test generation methods on a similar basis.

3.2 Initialisation and resetting

The simplest action to take to make a system testable is to ensure that the design is easily initialised to a predictable state. In the majority of cases this is necessary for normal operational reasons. However, there could be some 'don't care' states in the early stages of normal operation where a suitable initial value would aid testing.

Initialisation of a combinational circuit simply requires the application of an appropriate input vector and time to allow the signals to propagate through the network. Hence initialisation is only significant for sequential circuits.

The obvious means of initialisation of flip-flops is to provide an asynchronous clear, preset or both. This could require a very long wire and very heavy drive capability. However, only one initialisation connection is needed, or a small number in parallel to reduce driver loading.

In initialising a system, it is necessary primarily to check that the state of the system is as expected. However, the purpose of simulation is to find faults in a design. The designer must try to visualise what might happen if the design were faulty. For example, suppose it is required to reset some control latches.[1] If the clock input was inactive in the simulation then the correct things happen. However, in real hardware the clock might initialise to the active state, giving the wrong initial state to the system. The test program must check that the system initialises correctly independently of all other inputs which are regarded as 'don't cares' in a fault free system (see Section 10.1 point (a)). Thus a test must be run with the clock in the active state, or the design must be modified to ensure that that state cannot occur. Visualising such circumstances requires a lot of imagination and very careful consideration.

3.3   Scan design

A method which has been used successfully to ease the problem of testing, and which also solves the problem of initialisation, is known as **scan design**. In principle the design is divided into blocks of combinational logic separated by registers. The system is limited to synchronous designs,

Fig. 3.3. Principle of scan design.

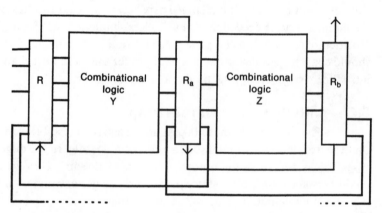

---

[1]   The term 'flip-flop' is frequently used of a two-state device which is triggered as a result of a clock edge ($1 \to 0$ or $0 \to 1$). The term 'latch' is used of a device which transfers data from data input to output when the 'clock' ('control' in this context) is in one state and blocks the transfer when the 'clock' is in the other state. This usage will be observed in this book.

therefore. Each of the registers can be configured either as a parallel load register for normal operation or as a shift register. In its basic form all the shift registers are connected end to end. Fig. 3.3 illustrates the idea.

To test a combinational logic block Z, say, a test vector is shifted to the appropriate part of the shift register, R_a at the input to block Z. This is **scan-in**. The signals propagate through Z and are parallel loaded into the register at the end of the block, R_b. The result can now be shifted out – **scan-out** and the appropriate bits checked against the expected result. Naturally there is a need to test the shift registers first.

This design technique has several advantages.

- Only combinational logic blocks have to be tested, which is a much easier problem than a general sequential network.
- The system is split into blocks, reducing the number of tests needed – see above.
- Several blocks can be tested at the same time. If R and R_a are built with edge triggered flip-flops then the output of block Y can be parallel loaded to R_a as the output from block Z is loaded to R_b. If suitable data was shifted into R as the previously mentioned data was shifted to R_a then both logic blocks are tested at the same time.
- Only two to four pins are needed for test purposes, namely, scan-in, scan-out and mode control (may be two). Indeed, scan-in and scan-out could be multiplexed. The advantage of using two pins is that a new test can be scanned in as the results of the old test are scanned out.
- The test structure – the registers – is part of the design in the case of synchronous designs. Although they are more complex due to the need for shift facilities, the equipment or silicon cost is relatively small.

Testing of the shift registers themselves can be done by scanning in a pattern and then scanning out the result. If the shift register chain is broken then the output will be all *1*s or all *0*s. The position of the break will have to be determined with the help of parallel loads. However, suppose that the control allows the register to shift in either direction. Suppose a fault equivalent to a s-a-*0* occurs at the sixth bit of a 10-bit register as shown in Fig. 3.4. If a pattern of ten *1*s is scanned in and then read out in the reverse direction, the expected pattern is ten *1*s. With this fault, the pattern read out is six *1*s and four *0*s. The position of the fault is where the pattern changes from *1*s to *0*s, namely, at bit six. S-a-*1* faults can be found in a similar manner with an input pattern of *0*s.

Having established that a path through the shift register exists, a test sequence of *00110011*... will exercise each flip-flop through all combinations of present and future states (present state *0*, future state *0* or *1*; present state *1*, future state *0* or *1*).

There is no restriction on the type of flip-flops used. However, IBM, in particular, have used latches. This makes it possible to test the clocking mechanism more thoroughly. This variation of the approach is known as level sensitive scan design – LSSD. A second property of LSSD circuits is that the steady state response (a 'long time' after the clock change) is independent of the circuit delays.

Random access scan is a technique where every flip-flop (or latch) is addressable. This requires each flip-flop to have its own address, and hence an address decoder is required, and a separate wire for each flip-flop. It is therefore costly in logic, pins and wiring space. The last two are frequently the most important for compact designs.

The primary advantages of scan design are access to internal ports of the design and a reduction in the number of test vectors and test time required. However, the scan-in/scan-out of test vectors is a serial process and hence time consuming. This could be offset by splitting the single long register into several shorter parts. This would then require register selection signals and hence pins. Pins are frequently a limitation of a design, so an increasing pin requirement is undesirable. A method of connecting test points on a chip is described in Gheewala (1989). Dual function pins are also becoming common, with the function being selected for many pins by one (test) signal.

The amount of logic – silicon area – required for test purposes will vary according to how many of the registers would be present without the scan design. Estimates vary with the design, up to about 20%. Dervisoglu (1990) gives a real example at 10%.

The scan system also prevents testing at full speed. This also applies to many other techniques. Testing at full speed gives rise to a completely different set of problems.

Finally, as with all testing methods, restrictions are placed on the freedom of the designer. If a design is to be reliable, serviceable and to

Fig. 3.4. Testing a scan register.

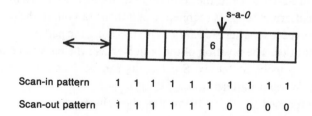

Scan-in pattern   1 1 1 1 1 1 1 1 1 1

Scan-out pattern   1 1 1 1 1 1 0 0 0 0

perform well, such restrictions are essential. One must learn to distinguish between freedom and licence – and not only in the electronic design field! Furthermore, it has been said that a chip can be designed and tested reliably without scan techniques, but the chip designer must remember that the chip will be built into a system and the system must be tested. A scan register on the chip will assist that even when not really required for testing the chip itself. This is of particular relevance in relation to the boundary scan technique.

3.4 Boundary scan

In recent years the problem of access to points internal to a system for test purposes has been the subject of considerable discussion. On PCBs there is a question of how to find a faulty IC. In late 1985 a joint test action group (JTAG) was set up to define a standard test interface. Led by European companies, others have since joined, resulting in IEEE standard 1149.1 (Evanczuk 1989).

The test architecture proposed is a generalisation of the scan technique. In essence a scan register 'surrounds' the board or chip (hence 'boundary' scan). In test mode the register outputs feed various points within the system to apply data more directly than could be done from the system primary inputs. Simularly, various 'test points' within the system can be fed as appropriate to the scan register to be read without having to control paths to the primary outputs (see Section 3.6). Although described as boundary scan, the scan register need not be physically on the boundary of a unit. Indeed, for speed there may be several scan registers in parallel.

The intention is that components conforming to the standard can be interlinked regardless of manufacturer. By suitable design, the scan chains can be around sub-circuits within ICs and then linked to other ICs or even between boards. The scan chains, or parts of them, are permitted to have LFSR (Section 3.5) and signature compression facilities to allow for testing of sub-systems. When the faulty component is identified, patterns entered by pure scan can isolate the specific fault. Several parallel scan registers are permitted. All four (or five) pins used by the test system are dedicated so that data can be scanned in or out while the logic is operating normally.

Fig. 3.5 shows the basic architecture of IEEE 1149.1. There are four or five terminals known as the test access port (TAP), a TAP controller, an instruction register and one or more data registers in parallel. All the registers are loaded serially from the input data terminal TDI and may be read out via the terminal TDO. The controller is a state machine in which most states can lead to one or other of two new states under the control of the TAP signal TMS. Fig. 3.6 shows a fragment of the state diagram. The

numbers *0* and *1* are values on TMS. If the system is in the reset state then it
will remain so while TMS remains *1*. If TMS goes to *0* then the state will
change to run-test/idle on the next rising edge of the clock, TCK. It will
remain in this state until TMS again becomes *1*. When TMS has changed,
the next rising edge of TCK moves the state to selecting the data register.
However, the data register scan does not begin until the next clock, and
then only if TMS has changed back to *0*. If TMS has remained at *1* then the
state moves to selecting the instruction register. Instruction register scan
begins on the next clock provided TMS has changed to *0*. If it has not then
the system returns to the reset state. The optional TAP signal TRST can
force an asynchronous reset. If this is not implemented (to save a pin) then
the reset state can be recovered in five clocks or less.

The test system may have many instructions, some of which are *private*

Fig. 3.5. Test access port (TAP) architecture.

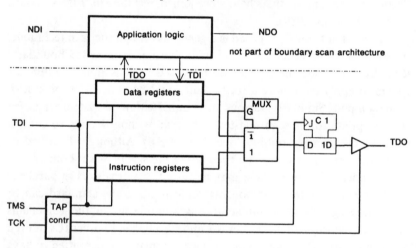

Fig. 3.6 Fragment of TAP contorl state machine.

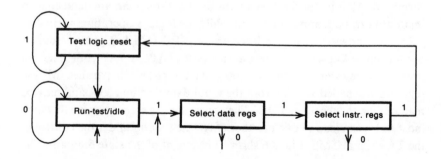

and will not be specified on the data sheet, and some of which are *public*. A number of public instructions have been defined but only three are mandatory. These are

- bypass – data passes from TDI to TDO in one shift register cycle, bypassing the (possibly) long data register. The instruction code is all *1*s,
- extest – external circuitry can be tested,
- sample – allows normal operation of the system with the ability to sample signals without affecting system operation.

Fig. 3.7 shows the data register organisation (Whetsel 1988). Three registers are shown. The scan bypass register is one bit and has been mentioned already. The device identification register is optional. It contains a compressed form of the JEDEC number, manufacturer's code etc. If this is not implemented, selection of this register gives the bypass register.

The boundary register can, in fact, be several registers in parallel, the instruction selecting the appropriate clock with the address as shown in Fig. 3.7. Fig. 3.8 shows one implementation of one such register. The register consists of a set of flip-flops, Q, connected as a shift register. Associated with each flip-flop is a latch, L. A multiplexer, A, selects from a normal (system) data input, NDI, or the previous flip-flop in the shift register to feed Q. A second multiplexer, B, in each cell selects from L or the NDI to feed the system logic. At the output of the system logic a similar arrangement applies, with the system logic output replacing NDI and the

Fig. 3.7. Boundary scan data registers.

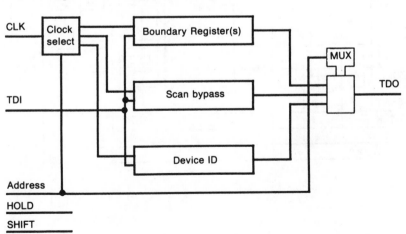

primary outputs, normal data output (NDO), replacing the logic input. The flip-flops are shown in one register, but it might be sensible to have one for the input pins and another for the output.

In normal operation system data passes from NDI to the system logic as fast as possible, only one multiplexer being involved (B). A similar arrangement applies at the output pins.

To shift test data into the system SHIFT is set. Data passes from TDI to the flip-flops in turn on the rising edge of CLK, a version of TCK selected under the control of data in the instruction register which chooses this data register (see also Fig. 3.7). HOLD is in the mode which prevents the data being passed to L. Notice that NDI can still pass to the system logic, and the system logic output can pass to NDO. Once all the test data has been shifted in, it can be transferred to the latches by changing the state of HOLD. All latches change together, so even if the B multiplexers are open there would be no rippling of data at the inputs of the system logic.

Parallel data capture can happen by setting SHIFT to control multiplexer A to receive data from NDI at the left of Fig. 3.8 or the system logic on the right of the diagram. CLK is allowed to go high. This data can now be scanned out. Of course, a new set of data can be scanned in at the same time.

Fig. 3.8. Boundary scan boundary registers.

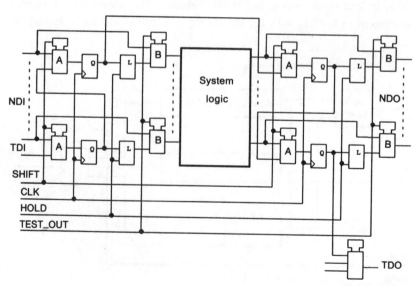

3.5   Self-testing

3.5.1   Dedicated test logic

A way to avoid having to provide pins on a chip or PCB for testing purposes is to make the logic self-testing. There are three approaches to this. The most obvious is to provide a module within the design which generates the test vectors, applies them to the logic and then checks the result, Fig. 3.9. This module must be self-checking.

Such logic need not take up a large amount of space – perhaps up to 25% of the total. The primary difficulty is that it is unique for each design and hence needs full design effort. A scan register, for example, can be a standard block which is the same for all designs. With self-testing one pin is still needed to indicate whether the system is passing the tests or not.

A major advantage of any self-testing approach is that the tests can be run regularly. For example

- at switch on,
- when the system is otherwise idle – not always possible,
- at specific time intervals.

This implies constant checking of the design and a rapid reporting of problems.

3.5.2   Signature analysis

The other two approaches to self-testing involve the use of random patterns applied to combinational logic blocks.

Suppose a set of test vectors is applied to a circuit under test. The signal at a particular node is fed to a linear feedback shift register (LFSR) whose initial contents are known – usually 0 (Fig. 3.10). As successive samples are clocked into the LFSR, the value it holds, the **signature**, depends on both

Fig. 3.9. Self-testing arrangement.

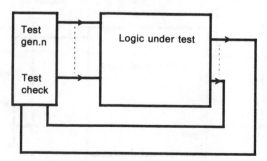

the incoming data and the previous contents. If the length of the LFSR is n bits it can be shown that the probability of a corrupt circuit producing the same resulting signature as a good circuit approaches 1 in 2^n as the number of samples clocked in exceeds n. Furthermore, the probability of two faulty circuits giving the same signature is small, so the value of the signature can be used to identify the fault causing failure as well as the fact of failure.

The problem here is in finding the signatures. To be a good test, a large number of test vectors must be applied. This will almost certainly be done on a simulator. If one wishes to be able to identify faults from the signature, then one also needs to simulate the full test for every possible fault and for many, if not all nodes in the circuit. Whilst this has to be done only once, it is extremely time consuming and expensive for any but small circuits.

3.5.3   Built-in logic block observation (BILBO)

The other disadvantage of signature analysis as described is that a separate LFSR and set of signatures is required for each node to be tested. If one wishes to test a large number of nodes, as is usual, this is prohibitive. A modification to the scheme is to use an LFSR to feed the primary inputs of a combinational logic block and a second one into which the primary outputs are loaded in parallel as shown in Fig. 3.11. The second register acts as a parallel input signature generator. This does not feed the inputs directly to

Fig. 3.10. Signature analysis.

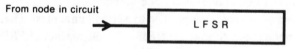

Fig. 3.11. Test of one block using parallel input LFSRs.

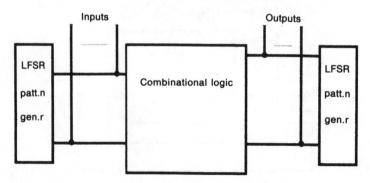

the LFSR feedback, but rather via a not equivalence gate, the other input of which comes from another flip-flop in the register as shown in the fragment in Fig. 3.12.

The value of random patterns for testing will be discussed further in the next chapter. In certain cases their use is valid. Here the input register generates a large number of input vectors for the logic under test and the output register generates a signature. If the two LFSRs can be switched into the scan mode then the LFSR seed can be scanned in and the signature can be scanned out and checked (unless it is checked on chip).

With many blocks of logic, such as is illustrated in Fig. 3.13, several logic blocks can be tested together. For example, blocks A and C can be tested at the same time using R_a and R_c as inputs and R_b and R_d as signature generators. In a second phase, block B is tested using R_b as input and R_c as output. A block D would be tested at the same time. Feedback between blocks is allowed only if the tests do not interfere.

This approach has the advantages of scan design and of reducing the size of the combinational logic blocks and hence test size. It generates its own test vectors rapidly, and so does not suffer the slow application of test vectors of normal scan testing. It is also possible to test the combinational logic at full operating speed.

Fig. 3.12. Parallel input LFSR

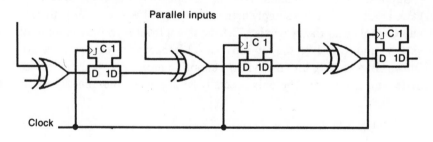

Fig. 3.13. Multiple logic block testing.

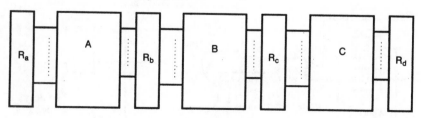

In practice the registers have four modes of operation as illustrated in Fig. 3.14.

$C1=0$, $C2=1$; Reset: all flip-flop D-inputs are 0.

$C1=C2=1$; Normal flip-flop: the parallel inputs are connected.

$C1=C2=0$; Scan path: one flip-flop connected to the next in shift register fashion.

$C1=1$, $C2=0$; LFSR: the parallel inputs are exclusive ORed with the data from an adjacent flip-flop.

3.6 Controllability and observability

3.6.1 Concepts

Having made some effort to design a system that is testable, it is desirable to have

- some measure of how successful the effort has been,
- some indication of areas of the system which are difficult to test, and hence areas where improvement might be made. In particular one needs to know what is actually not testable. Measures of testability can be used to guide test strategy, as will appear in Chapter 4.

Suppose there is a large block of logic as shown in Fig. 3.15. Only a few gates are shown. All lines imply many more gates. The curved lines from C and Q imply that these signals affect the signal from E to the top input of gate A1 and from P to the top input of A2 respectively through logic not shown. In a test rig internal points cannot be monitored (e.g. the device under test is a chip). The only access is via inputs to the unit known as

Fig. 3.14. One stage of a BILBO register.

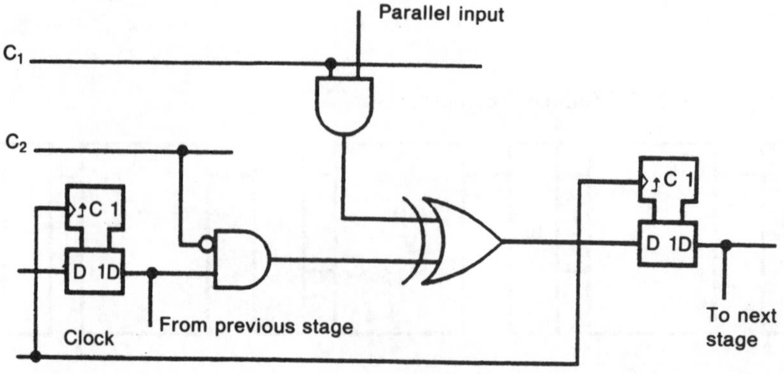

primary inputs (PIs) and via outputs of the unit, the **primary outputs (POs)**. Suppose there is a potential fault at point W. The procedure is as follows.

- Find a path from the PIs to the point W and an input pattern such that the signal at W in the fault free circuit is the *opposite* to that of the faulty circuit. Thus, if the fault at W is s-a-*0*, try to place a *1* at W. This is shown as the path B–E–W. Along the way other signals also need to be set properly to allow the signal from B to pass along this path. In particular, the signal from D must be such as to produce a *1* on the second input of the AND gate, A1, hence allowing the signal from E to pass through. Similarly the signal from A must cause a *0* at the input to the OR gate, thus allowing the signal from B to determine the value at E. C will also have to be set appropriately and many other similar combinations will be required. It is said that the signal at W is being **controlled** and its **controllability (CY)** is a measure of the difficulty of setting W to a known value.

- Find a path from the faulty point W to an output Z and an input pattern which is compatible with that in the previous paragraph such that the signal at Z under the condition of the fault at W is different to that under *the same input conditions* in the fault free circuit. The input conditions for a s-a-*0* at W are such as to try to place a *1* at W. If a *1* at W produces a *0* at Z for the fault free network, then the *0* at W produced by the fault must produce a *1* at Z for the fault to be **observable**. Again it will be necessary to set up conditions on the PIs so that the path is open. For example, the PIs P and Q must be such as to produce a *1* at the first input of the AND gate, A2. This is the process of making the fault observable at

Fig. 3.15. Example of controlling and observing internal signal X.

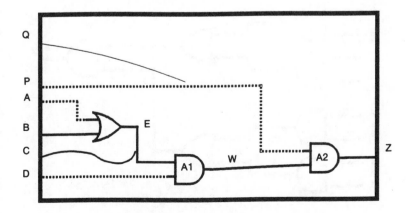

the POs, and a measure of the difficulty of doing so is called the **observability (OY)** of W.

It should be appreciated that, to control a particular node deep within a circuit, there may be many paths back to the PIs. As a simple example, consider a fault at W in Fig. 3.16. A path W–J–A leads to a primary input. For this to be used signal K must be *0* and hence so must C or D. C can be selected arbitrarily for this example.

Similarly, for W to be observable at Z, N must be *1*. This implies that M should be *1*, say, and hence that E and F must both be *1*. M was selected arbitrarily; L would have been equally good. It will be seen that there is an explosion of data, and the measurement of controllability and observability must take into account the ease or difficulty of setting all these signals. The selection of such signals is pursued further in Chapter 4.

Much can be done by a designer to ensure that all points in a block of logic are both controllable and observable. However, intuitive methods can miss particular cases and automatic methods of assessing how good a design is have been developed. These should be used whenever possible. They can identify particular points of difficulty for which redesign should be considered. Whilst this may not always be possible or economical, it does at least provide a warning. Regular use of such aids should also help users to develop a good design style.

3.6.2 Controllability

Consider a two input AND gate. The controllability of the output is a function of

Fig. 3.16. Illustration of data explosion finding CY and OY.

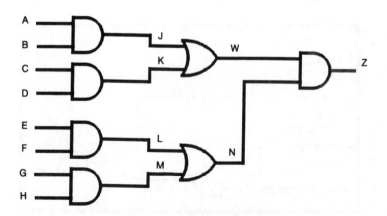

- the difficulty of creating a path from input to output, which is a function of the circuit – an AND gate,
- the controllability of the inputs.

One simple method of defining controllability for networks of simple gates (Goldstein 1980) is to assume the first factor is always 1, and to define the controllability of a node as the number of nodes that must be set in order that the node in question can be set to the required value. This applies only to combinational networks. Thus are defined

Combinational controllability 0 (CC0)
Combinational controllability 1 (CC1).

For the AND gate, any input set to *0* will cause the output to be *0*. Hence CC0 of the gate output is the minimum of the *0*-controllabilities of the gate inputs plus 1 – for the gate itself. To set a *1* to the output of the AND gate requires all inputs to be *1*. Hence CC1 is the sum of the *1*-controllabilities of the inputs plus 1. For an OR gate all inputs must be *0* to place a *0* at the output. Hence CC0 is defined to be the sum of the *0*-controllabilities of the inputs plus 1. Only one input need be a *1* to set a *1* at the output so CC1 is the minimum of the *1*-controllabilities of the inputs plus 1.

Consider now point W in Fig. 3.16. If W is to be *0*, J and K must both be *0*, and hence one of A and B and one of C and D must be *0*. The controllability of an input is defined to be 1. Thus

$$CC0(J,K) = 1(A,C) + 1(gate) = 2$$

and the controllability of W is

$$CC0(W) = 2(J) + 2(K) + 1(gate) = 5$$

For a *1* at W either J or K must be *1* and both of (A and B) or (C and D). Hence

$$CC1(J) = 1(A) + 1(B) + 1(gate) = 3$$
$$CC1(W) = 3(J) + 1(gate) = 4$$

Where sequential circuits are concerned, two further quantities are defined. These are the **sequential controllabilities SC0 and SC1**. A 'sequential node' is a circuit node which is held constant for one time period. If a node has to be held at *1* for five time slots in order to propagate a logic *0* to a given other node then the given node has SC0 of five. SC0 and SC1 are clearly zero at PIs and for any purely combinational circuit. For a sequential circuit, 1 is added to SC values but zero to CC values.

The values of controllability can, if required, be modified at each logic stage by a factor related to the function of the logic element. In the above for an N input AND gate, the value of CC1 is increased to the sum of CC1 of the inputs plus one. This is a measure of the gate function. On the other hand CC0 is only increased by 1 – again a measure of the ease by which a *0* is passed to the output compared with that of passing a *1*.

With this form of controllability, the values increase without limit. High values represent poor control. Bennetts (1984) prefers a normalised value between 0 and 1, 0 being uncontrollable and 1 being perfectly controllable (e.g. a primary input). He defines a **controllability transfer function, CTF,** for a logic element which is independent of the logic level (*0* or *1*). If $N(0)$ is the number of ways of producing a *0* on the output and $N(1)$ the number of ways of producing a *1*, then

$$CTF = 1 - \frac{N(0) - N(1)}{N(0) + N(1)}$$

Consider a three-input AND gate. There are eight possible input patterns of which one gives a *1* output and seven a *0* output. Hence

$$CTF = 1 - (7 - 1)/8 = 0.25$$

The question now arises as to what function of the input controllabilities should be multiplied by the CTF. Bennetts considers the geometric and arithmetic means.

$$\text{geo. mean} = (CY_{X1} * CY_{X2} \ldots * CY_{Xn})^{1/n}$$
$$\text{arith. mean} = (CY_{X1} + CY_{X2} \ldots + CY_{Xn})/n$$

Consider Fig. 3.17. Input A is tied to a logic level in order to use a spare two-input gate in an otherwise partly used IC, rather than an extra IC of buffers. A has zero controllability since the primary inputs of the network can never affect it. Hence the geometric mean function would be

Fig. 3.17. Zero input controllability.

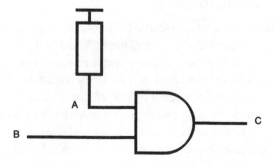

$$CY_C = (0 * CY_B)^{1/2} = 0$$

which is clearly wrong. It should be

$$CY_C = CTF_\& * CY_B$$

The arithmetic mean gives

$$CY_C = (0 + CY_B)/2$$

which is not zero, but division by 2 ('n') is still incorrect.

Both problems could be solved by preprocessing the network to find situations like this. The CTFs are calculated once and held in a library, as are the functions for calculating CY. For the case shown in Fig. 3.17, the CTF and the function used should be that for a one-input gate rather than a two-input gate. Thus, in the geometric mean case, the uncontrollable input does not feature, and 'n' is 1 rather than 2 in both cases. Bennetts prefers the arithmetic mean calculation with this modification. Notice that Goldstein's controllability measures get this right – the tied off input adds nothing to the controllability.

A second problem arises with a device such as a D-flip-flop with asynchronous inputs as shown in Fig. 3.18 (for specification see Section 1.5). Quite clearly, if the preset or clear or both are set to 0, Q is not controllable. Furthermore, if the preset or clear are to be set to 0 then the controllability of D and clock are irrelevant. On the other hand, when both preset and clear are 1, the controllability of Q is a function *only* of the controllability of D and clock. Thus, a useful way of defining CY_Q might be

$$CY_Q = CTF * (CY_{clr} + CY_{pr} + CY_D * CY_{clk})/3$$

However, if the preset and clear are tied off, this expression should be

$$CY_Q = CTF * CY_D * CY_{clk}$$

Fig. 3.18. D-flip-flop with asynchronous inputs.

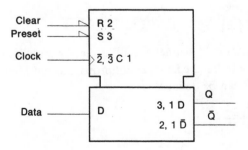

and not one third of this as would be obtained from the previous expression with $CY_{clr} = CY_{pr} = 0$.

3.6.3   Observability

Observability is the ease with which a signal at a particular node can be observed at a primary output. In general, to reach that output it must pass through several circuit elements as shown in Fig. 3.15. For a given circuit element, the observability at the output is a function of

- the circuit element,
- the observability of the inputs on the path to the primary output,
- the *controllability* of all other inputs affecting transfer of this signal along the path to the primary output.

Referring to Fig. 3.16, the observability of W at Z is a function of the observability at W (1 in this case as W is the signal of interest) and the controllability of N, the signal which must be a *1* to allow the faulty signal W to propagate to the output, Z.

A measure of observability corresponding to the combinational and sequential controllabilities is constituted by the combinational and sequential observabilities CO and SO. These are the minimum number of combinational or sequential nodes respectively which must be set to enable a fault to pass from its source to a primary output.

For signal W in Fig. 3.16 the combinational observability is derived as follows. The observability of an output is 0. The observability of the input of an AND gate (W) is given by

$$CO(W) = CO(Z) + CC1(N) + 1$$

This takes into account the fact that N must be *1* to allow W to be observable at Z.

$$CC1(N) = CC1(L) + 1$$
$$= (CC1(E) + CC1(F) + 1) + 1$$
$$= 4$$

Hence

$$CO(W) = 0 + 4 + 1 = 5$$

The factor corresponding to Bennetts' CTF is the **observability transfer factor**, **OTF**. This must be *0* if the fault cannot propagate and *1* if it always propagates. The OTF can be defined as

$$OTF = \frac{N_{pass}}{N_{pass} + N_{block}}$$

where N_{pass} is the number of ways a fault can pass from its input to its output and N_{block} is the number of ways it can be blocked. For the three-input AND gate there is only one way for a fault on a particular input to pass to its output – all other inputs are 1. However, there are three ways in which it can be blocked – either or both of the *other two* inputs being 0. Thus

$$OTF = 1/(1+3) = 0.25$$

OTF values, like CTF values, can be held in a library.

Calculation of observabilities should start at the primary outputs where the observability is 1 (Goldstein: $CO = SO = 0$). Using the values of the product of OTF and supporting controllabilities, the observability of each driving input is computed. This is now repeated back through the network. In a single pass per primary output, the observability of all nodes is computed. If one starts at a faulty node then a computation from node to output has to be carried out for every node – a lot more computation.

3.6.4 Testability

The testability is a measure of how controllable and observable a circuit is. It is a function of both controllability and observability. In Goldstein's proposal using CC0, CC1, etc., a suitable weighted sum of the six values is used. In Bennetts proposal

$$TY_i = CY_i * OY_i$$

where TY_i is the testability of node i.

$$TY = 0 \text{ if } CY = 0 \quad \text{ or } OY = 0$$
$$TY = 1 \text{ if } CY = 1 \quad \text{ and } OY = 1$$

Also if CY and OY are both 0.5 (say), $TY = 0.25$. In other words, if it is 50% more difficult both to control and to observe a node then it is 75% more difficult to test it.

The testability of a design is defined by Bennetts to be

$$\sum_i^n \frac{TY_i}{n}$$

where n is the number of nodes.

To use a testability analysis, one might do two things. First, plot a histogram of node testability against number of nodes with that testability. Separate histograms of controllability and observability may also prove to be useful. This will indicate if any nodes have particularly poor testability. If the controllability is bad then some effort may be made by designing some more direct control in the area of this node or the path leading to it. This can be done with multiplexers as shown in Fig. 3.19. Suppose the node W in the original design (a) is found difficult to control. A multiplexer is inserted into the network somewhere before W. In normal operation the signal C is *1*, allowing the function of the network to proceed normally. In testing, when node W must be controlled, C is set to *0* and P used as data to set the required value of W. Insertion of this multiplexer will also improve the controllability of signals which are affected by W, such as Y.

Poor observability is improved by designing easier access from the signal to a primary output. This could be done by providing additional outputs (test points) or by providing a multiplexer between an internal signal and

Fig. 3.19. Improving controllability.

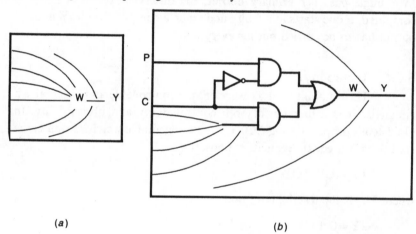

(a)   (b)

Fig. 3.20. Improving observability.

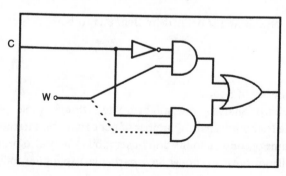

an existing PO as shown in Fig. 3.20. This multiplexer effectively bypasses a long chain of logic as represented by the dotted line. Again, this multiplexer will improve the observability of the signals controlling W.

Of course, the multiplexers will reduce the speed of the system under normal operation. It follows that the choice of position to place multiplexers is critical in order to improve the controllability and observability of the largest possible section of difficult to control/observe logic, without at the same time affecting the speed in too adverse a manner.

As an example of a sequential circuit, consider a long counter as shown in Fig. 3.21(a). To control W requires 256 clocks. A multiplexer before A reduces this to 16 clocks. A multiplexer just before W makes it fully controllable, Fig. 3.21(b). The former still requires 256 clocks to control Z. The latter leaves 16 clocks to control Z, but 256 clocks to control B. It might be useful to put in both multiplexers.

There are numerous additional matters that are required for full consideration of this subject. It is hoped that this account will serve as a useful introduction. There appear to be few recent references in this area.

Fig. 3.21. Controllability of a sequential circuit.

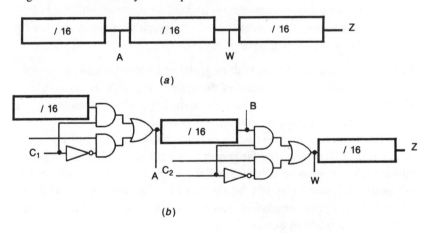

(a)

(b)

4

Exercising the design in simulation and test

4.1 Objectives and approaches

4.1.1 Objective

The first objective of any set of tests must be to show that the system does what it was designed to do. This means

- does it perform the necessary functions correctly?
- does it perform its function within the required time specification?
- are there any circumstances, normal or unusual, under which it can get into a forbidden state from which it can only recover by drastic action (e.g. system reset)?

The above presumes that any fault or group of faults which could possibly occur would affect the operation of the system. If a fault or faults do not affect the operation then there must be some logic which is redundant. *It will be assumed that none of the logic is redundant.*

On the further *assumption that the design is good*, a second objective is to be capable of detecting any fault or group of faults within that system. However, there is some debate as to the level of detail into which it is necessary to go and this will be the subject of further discussion. The distinction between simulation testing and testing the manufactured hardware was made in Section 3.1.2.

4.1.2 Modelling faults

Faults in the design, as opposed to those which occur in production, cannot be 'modelled' in the usual sense of the word. A network description is entered into the simulator and this is a model of the network. If, as a result of simulation, outputs are obtained which are different from those expected from considering the specification, then the network as described to the simulator is faulty. In that sense the fault model is a faulty description of the network. A list of what might more usually be called fault models was given in Section 3.1.2, which refers to both design and manufacturing faults.

For production testing the design is presumed good. The question arises as to what faults can occur, and how can they be represented in the simulator in order to generate a set of tests which will indicate that the system is faulty? This set of tests must give at least one output which is different from the output which would be obtained from a fault free system with the same inputs.

An obvious example of such a fault is a wire being broken, perhaps due to rather thin aluminium going over a 'step' in an IC or over-etching on a PCB. In a TTL circuit, the inputs are connected via a diode and a resistor to the positive supply as illustrated by the input circuit fragment of Fig. 4.1. Hence an open circuit input is equivalent to the higher logical voltage level. This input can be described as 'stuck at *1*.' Another possible fault is a wire shorted to a power supply or ground giving a stuck at *1* or stuck at *0* respectively. On an IC there are other types of fault which could have other effects. These include dislocations in the silicon crystal, dirt in the processing etc. The electrical effect of these must be interpreted in logical terms in each case. This is known as **fault modelling**. Gheewala (1989) has identified 26 possible faults in a two-input CMOS gate and 150 for a flip-flop. It is not said how many can be represented as s-a faults.

The conventional approach to testing is to model all faults as stuck at *1* or stuck at *0* at device inputs and outputs. This implies that any fault internal to a gate can be represented as one or other of these two faults. It is not too difficult to show that this is not so. For example, suppose there is a break at the collector of a TTL circuit so that it can be represented as

Fig. 4.1. Input of a TTL circuit.

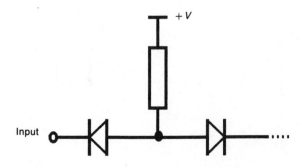

shown in Fig. 4.2. The transistor can pull the output down when switched on. When switched off, the TTL circuit will not pull the output up, but the input to the driven circuit (e.g. Fig. 4.1) will. This is neither stuck at *0* nor stuck at *1*. The rise of the output may be slow and show up as a delay fault (Section 9.8.1). The designer is warned!

For MOS circuits, some internal faults require the gate to be treated as a sequential circuit. These are due to a transistor stuck open and the very high input impedance of the following MOS device (see Section 4.6). The approach to testing also usually assumes single faults and hopes that multiple faults will not mask all of the possible single stuck at faults and hence will be detected. It has been calculated (Agarwal 1981) that a circuit of 1000 nodes and multiple faults at up to six places has 10^{17} different faults. The same author claims that a test set giving 100% coverage on single faults will detect 98% of the multiple faults at up to six places for circuits where the fan-out of each element is limited to one place. However, the cover drops rapidly in *real* circuits which have higher fan-outs.

Another author (Schultz *et al.* 1988) claims that a test set that finds 100% of single faults will also detect all multiple faults provided there are restrictions on the connection method. This includes no reconvergent fan-out, so systems using exclusive ORs are not allowed!

Schultz also reports that if two faults are found by only one test each, the double fault involving these two faults is often undetectable. Thus a search for faults found by only one test is made, and these specific vectors

Fig. 4.2. Internal fault in TTL gate is not stuck-at fault.

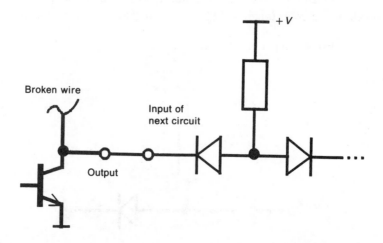

simulated for double/multiple faults to check if there is any fault masking. Where there is, new tests for these multiple faults are generated. This would seem to be a useful addition to the test programmer's armoury.

Practice shows that the assumption of a single s-a fault usually works out reasonably well. The assumptions of Agarwal (fan-out = 1) and Schultz (no reconvergent fan-out) are not reasonable, but their work warns of the limitations of the single s-a assumption.

Finally, faults resulting from two or more wires stuck together but not s-a-*0* or s-a-*1*, *bridging faults*, are usually ignored, probably because their analysis and detection are very difficult. In fact, tests for s-a faults do not find these faults. Some *ad hoc* methods can help (Section 4.2).

4.1.3 Assessment of test coverage

The question that now arises is that of how to assess the value of a set of tests. The first step is to confirm that the specification is met, with several (numerous?) sets of data for each operation where relevant. The number of tests and hence confidence in the design must be balanced against the cost of additional testing. The value of tests for a particular purpose is somewhat subjective, and must be agreed between designer and customer at the time the system is specified.

Assessment of the tests to find faults due to manufacturing defects is generally done by running a simulation of the system without faults (fault free) and then again with a fault introduced to see if the simulation results are different from those of the good circuit. The number of tests is so large that the designer could not possibly work out all the results for the fault free circuit. Hence there must be some measure by which the fault free circuit can be said to meet the system specification before the fault free simulation is run. That was the purpose of the tests of the previous paragraph. The simulation is then repeated for all faults which the tests are designed to find. The proportion of faults detectable is known as the **fault cover**.

The number of possible faults is very large, and hence the necessary simulation time is also very large. Note that these simulations are performed to assess the value of the test program, not to prove the excellence of the design. Much effort has been put into developing methods of assessing fault coverage and this is the subject of Chapter 9.

4.2 Testing for design faults

This section is concerned with the generation of tests to prove that the system that has been designed meets the specification. It is not concerned with testing for manufacturing faults, though the tests produced are often useful as a starting point for that purpose. There is no absolutely 'right' way to do this, but application of 'common' sense will be of great help.

Consider, first, an arithmetic logic unit such as the 74181 or equivalent devices. This is a purely combinational logic block, so has no problems of clocking. The chip is shown in Fig. 4.3. It performs 32 functions selected by four S and one mode bits. Each function operates on two 4-bit operands, A and B, producing a 4-bit output, F. For the arithmetic functions there is also a carry_in data signal input and a carry_out data output. Finally, there are two output signals, G and P, which are used in conjunction with the 74179 chip to build fast multibit adder/subtracters (G stands for carry generate, P for carry propagate). In fact, many of the logic functions are of no interest. The chip was designed to execute about 16 real functions with simple control. The other functions are what happens 'by accident.' However, all 32 functions are specified on the data sheet and so must be tested.

First of all, try to set all the outputs to 0s, then to 1s *and then back to 0*s. To do this, select a simple function such as 'copy A to F.' Three input vectors are needed.

$$A = 0101 \qquad A = 1010 \qquad A = 0101$$

The control and mode bits are set appropriately. Carry_in should be irrelevant. If the values of F produced are copies of A then we have proved the following.

- All outputs can be set to 0.
- All outputs can be set to 1.

Fig. 4.3. 74xx181 ALU.

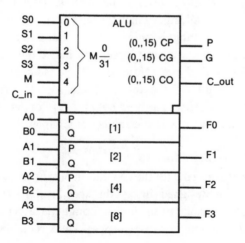

- All outputs can be changed from *0* to *1* and vice versa. Hence also there is no immediately obvious storage within the chip – possibly due to faulty connections. In a simulation the 'faulty connections' may be due solely to mistyping the circuit data.
- One function works – copy A to F.
- Supposing that the layout of the chip is linear in the order of bits, adjacent bits are not stuck together (if the layout assumption is not true, some other appropriate pattern should be chosen). Note that input patterns *0000, 1111, 0000* would not have detected this. Note also that the input patterns are dependent on the layout of the chip. The chances of other pairs of bits being stuck together is remote if none of these pairs are. During the design, the simulation needs to use other patterns to show correct data entry.

It is now necessary to test all 32 functions in turn. Tests of the purely logic functions should be done with several sets of operands and with the carry_in varying in value to give confidence that the functions are, indeed, independent of it. Similarly, one operand functions (such as copy A) should be done with several values of the other operand (B in the example) to show its independence of that operand. It is the tests with varying C_in and B in this example which are frequently omitted and production of which, therefore, need careful thought by the designer.

The three tests above will not test the carry chain nor the G and P signals. To do this, the A and B input patterns for tests on the add and subtract functions should be chosen to show that these two signals work correctly on several input patterns. Fig. 4.4 shows the G circuit. $G_i = A_i$ & B_i; $P_i = A_i + B_i$ where i is the bit number (0 to 3). Operands should be chosen so as to set G via each gate separately and in combinations.

Fig. 4.4. Generation of the 'G' signal of the 74181.

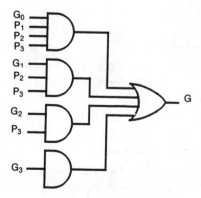

For the arithmetic operations there are a very large number of paths through the chip since carries can propagate by a number of different routes. A careful study of the logic should show how to ensure that a $0–1–0$ (or $1–0–1$) pattern goes through most gates. Another important point is that some faults, even in purely combinational logic, may well give an output dependent on a previous value. This is particularly so in the design phase where a wrong signal may have been specified accidentally. By ensuring two changes rather than just the one, many such faults will be found.

For this particular logic element it is just about possible to apply all possible input patterns – 16K – though not all sequences. However, this chip only contains about 63 gates. For many thousands of gates and several tens of inputs, applying all possible patterns is not possible. The author estimates that a manually generated test set consists of three to six operands for each function and perhaps 20 to 30 for *one* arithmetic function – or shared by several. This gives less than 200 test vectors.

A point not well illustrated by this example is the need to ensure that 'special cases' are properly designed. For example, the designer of an arithmetic unit will want to check extreme cases – what happens when a result is too large to be represented, is the overflow set (there isn't one in this example), what is the result of divide by zero? . . . In this example, the addition of two operands giving a sum greater than 15 should be tried. The result of not doing this can be disastrous. An anecdote was related in Section 2.1.5 explaining why.

Although not stated, many of the tests on an ALU can be designed to test for several faults at the same time. For example, the tests of the function 'copy A to F' tested four nominally independent routes together. As a further example, consider the structure in Fig. 4.5. This represents part of the data path of a system. The path is 32 bits wide and in the section shown there is no interaction between the bits. Hence, with one test vector of 97 bits (32 x, 32 y, 32 z and enable), it is possible to test for a s-a-1 at the outputs of all the exclusive ORs. Another vector tests for the s-a-0 and so on for faults which are not equivalent. Thus each vector tests for 32 faults. This is functional testing with an eye on hardware testing. Without special help an automatic test generator would find some of these, but would be fortunate to find them all. As before, if the layout is considered the test patterns can be designed to check for many bridging faults.

For sequential circuits the problems are more difficult. Consider, for example, a counter. It may be possible to parallel load an initial value, and it is necessary to check maximum and minimum value indicators. A way of making this more testable was shown in Fig. 3.21. Again, an understanding of the structure of the circuit will result in functional checks giving a high fault coverage for relatively few tests, but here many of the tests will involve several test vectors in sequence.

An interesting example from the author's experience is the testing of the piece of pipelined logic shown in Fig. 4.6. Access for testing purposes was through inputs, outputs and clock only. The logic was 3-bit wide arithmetic and the total number of gates was about 180 (a 1976 ECL gate array). The primary functions of the unit were tested with only 19 test vectors.

4.3 Testing for manufacturing faults

The development of tests to prove that the design meets its specification is somewhat *ad hoc*. It requires a lot of *very careful thought*. The test patterns and results must be meaningful to the designer(s). Testing for manufacturing faults is different. The test patterns need not be

Fig. 4.5. Part of a system data path.

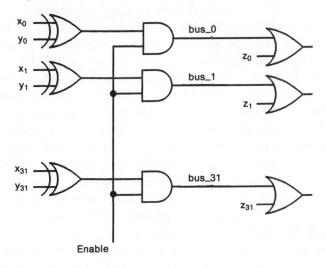

Fig. 4.6. A sequential logic example.

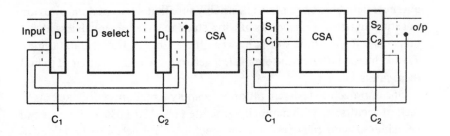

meaningful. However, to check whether faults are detected, it is necessary to begin with a fault free network. The approach of Section 4.2 is designed to determine this.

Tests for manufacturing faults can be generated automatically by systematic methods. They will find any fault if a test for that fault exists. Most of the methods of test generation identify a fault for detection and then generate a test to detect it by algorithmic methods. This and following sections are concerned with how tests may be generated to find a possible fault in the *manufacture* of a design *which is known to be good*.

Generating a full set of tests can take a lot of resources. The tests as generated in the previous section already exist. In the example of Fig. 4.6 the 19 vectors were found to give a s-a fault coverage of around 90% which included a separate vector for each clock change. This indicates that these tests are a very good starting point for manufacturing test generation.

In the example of Fig. 4.6 the registers could be regarded as several shift registers in parallel, with access at serial input and output points. The fault simulation (Chapter 9) indicated which faults were not checked. Additional vectors were generated (manually) to test these. In several cases a number of input vectors had to be applied just to set up the internal states for the required test. These last 10% of possible faults required over 100 vectors! This graphically illustrates the difficulty of obtaining 100% fault coverage. The example also shows the great effort needed to produce the additional tests, and shows why automatic means to generate them are essential. It is not clear what value the extra 100 test vectors had in terms of assessing the functionality of the logic – no extra design faults were found. No attempt was made to redesign the circuit to be more testable.

It is found in practice that a relatively small number of hand-generated tests will find a relatively large proportion of faults. The experience of several authors suggests that a 70% fault coverage can be obtained for a quite tiny set of test vectors, as was also indicated above. The effort expended by the designer is quite small, and it leaves the automatic system to generate tests only for those faults which would take up a lot of manual effort. Furthermore, if the structure of a diagram such as Fig. 4.5 reflects the physical layout of the final product, then by testing with 0s and 1s in alternate data bits, one can be reasonably sure that adjacent data wires are not stuck together. Other pairs of wires stuck together are much less likely, other than for multiple shorts between the enable and several data lines. The likelihood is that an automatic test generator would not attempt to find these faults at all.

One approach to test generation is to apply a random set of patterns in the test vectors. As with manually generated tests, 70–90% fault cover can be achieved with relatively few test vectors. Care needs to be taken to

Table 4.1. *Cubes of a two-input AND gate*

A	B	G
1	1	1
0	X	0
X	0	0
D	1	D
1	D	D

control clock signals, and, possibly, some control inputs by non-random inputs. This should lead to better fault coverage. For example, a reset input should not be applied in 50% of test vectors as would happen with truly random test patterns. These tests are unstructured by nature, and are of little use for determining whether the design is correct. The technique is again useful only for tests to check that a known good design has been properly manufactured.

It has been reported that certain logical structures are resistant to random pattern tests. This simply means that the fault cover for a given number of randomly generated test vectors is disappointingly low. In one example 12K patterns gave only 77% cover (Wunderlich 1987). The provisions of the previous paragraph will help this problem but not completely solve it. An alternative is to use a more sophisticated process of selection for the patterns to use. In the case quoted, this improved the coverage of the 12K patterns to 99.7%.

4.4 The D-algorithm

4.4.1 Basic ideas

In 1966 J. P. Roth of IBM published an algorithm by which tests could be generated for any fault in a combinational network if such a test existed (Roth 1966). Clearly these are also likely to be unstructured and hence not much use for checking the correctness of the design without considerable arrangement by the designer. It may be used to generate all patterns for production testing. However, as has been indicated earlier, the manual approach will probably find a large number of faults for relatively few vectors.

Since 1966 there have been many other papers published on this matter. By far the greater number of them are derived from Roth's algorithm. An understanding of the technique is essential for appreciation of later work.

Consider, first, an AND gate with inputs A and B and output G. The operation can be described by the truth table shown in Table 4.1 or geometrically as shown in Fig. 4.7. Each line of a full table is seen to be a **vertex** of the **cube**. The line *0 X 0* represents *two* vertices – *0 0 0* and *0 1 0*. The values of *D* may take one of two values, *0* and *1*. The line *D 1 D* implies that if A is *1* then so is G and if A is *0* then so is G.

The distinction between *X* and *D* is as follows. Suppose we wrote a vector of signals as *X X 1*. Each value of *X* is independent. Thus, if the first *X* is *1*, the second could be *0*. This is *not* true of the vector *D D 1*. In this case both values of *D* must be *the same*; either *1* or *0*. Thus while *X X 1* represents four vertices of the cube, namely, *0 0 1, 0 1 1, 1 0 1* and *1 1 1, D D 1* represents only two vertices, namely *0 0 1* and *1 1 1*.

Returning, now, to the AND gate, the vertex *D 1 D* represents four tests. Suppose we wish to test for input A s-a-*1*. To do so we need to try to set this input to *0*. B is set to *1* as in the vertex *D 1 D*. In the good circuit the *0* on A will cause a *0* on G. However, in the faulty circuit A is s-a-*1*, so G will be *1*. Thus A is *0* in the good circuit and *1* in the faulty circuit. Similarly, to test for A s-a-*0*, attempt to set A to *1*. G will be *1* in the good circuit and *0* in the faulty one. In each case G takes *the same* value as that to which we try to set A.

Note now that, if G were s-a-*1*, the effect would be the same as if A were s-a-*1* and B were *1*. Similarly for G s-a-*0*. The vertex *1 D D* then caters for the case where B is stuck. A is set to *1* to ensure that the fault propagates to G.

Consider, now, a NAND gate with inputs A and B. The two D-cubes corresponding to those of the AND gate are written (*D 1 D̄* or *D̄ 1 D*) and (*1 D D̄* or *1 D̄ D*). The first of these implies that if B is *1* and we wish to test for A s-a-*0*, then we try to place a *1* on A. The result in the good circuit is a *0* on G – that is, the signal is of different value to that on A. It is sensible to

Fig. 4.7. Geometrical representation of AND gate operation.

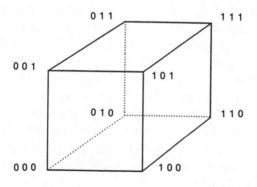

choose the first of each pair or the second of each pair, but it is not essential to do so.

Suppose, now, that we have a test vector on the primary inputs of a circuit. Consider a given signal S somewhere within it.

- If S is *1* for both the fault free and faulty circuits then the D-cube has this signal as a *1*. Similarly if both circuits have S at *0*.
- If the fault free circuit has S = *1* and the faulty circuit S = *0* then in the D-cube S is represented as *D*.
- If the fault free circuit has S = *0* and the faulty circuit has S = *1* then the D-cube represents S as *D̄*.

These assignments are more restrictive than those suggested earlier and have been adopted as a convention. They are not essential to the use of the technique. For this assignment to be always possible the circuit must not contain feedback.

It should be noted that the D-cube produced by these considerations can now be used as a definition of the test on the primary inputs. If a D-cube represents a test then there must be a path from the point of failure of the faulty circuit to a primary output which contains only *D*s and *D̄*s.

Three questions arise.

- How can the sets of signals on a logic element be determined to test for a fault? These sets of signals are known as **primitive D-cubes of failure**.
- How can a path to an output be found, and how can the signals controlling this path be set from the primary inputs? A hint of the answer to this was given in Section 3.6.1 in relation to controllability and observability.
- How can the primary inputs be set to produce a signal in the fault free circuit which is opposite to that produced by the fault?

To detect a fault it is necessary to perform the following.

(a) Find input patterns for which the good circuit, G, has a *0* output and the faulty circuit, F, a *1* output, or vice versa. Then find compatible inputs for G and F. Thus, referring to Fig. 4.8, if G = *1* and F = *0*, find values of A which are both *1*, both *0* or are *1X* or *X1*, and similarly for the other three inputs.

(b) Having found the required inputs to this sub-circuit, trace back to the primary inputs to find how they should be set to achieve the required values of A, B, C and E.

(c) For this case, set the output of the sub-circuit to D, and trace forward to an output. At each logical element passed, set the other inputs so that D is transmitted and trace these inputs back to the primary inputs.

4.4.2   Primitive D-cubes of failure

This section deals with point (a) above. An important point to make is that the *type of fault* that can be detected is not specified. The stuck at fault on an AND gate is by no means the easiest to understand. Consider the circuit shown in Fig. 4.8. Fig. 4.8(*a*) shows the fault free circuit in which

$$G = A.B + C.E$$

The faulty circuit has the logical function

$$F = A.B + E$$

This might be due to C being stuck at *1*, but need not be so. However, it can be said that the faulty circuit is equivalent to a good circuit in which the values of C is *1*.

Write down two tables.

- In the first, write all the circumstances under which the *fault free* circuit, G, gives a *1* output and the *faulty* circuit, F, gives a *0*. This is shown in Table 4.2(*a*). The lines of this table represent the vertices of a five-dimensional cube – hence the references in the literature to D-cubes even when not 'three-dimensional'.
- In the second, Table 4.2(*b*), write down the circumstances in which the *fault free* circuit gives a *0* output and the *faulty* circuit a *1* output.

Considering Table 4.2(*a*), the *fault free* circuit gives a *1* output if either A AND B are *1* or if C AND E are *1*. In the first case the values of C and E are immaterial. They could be *1 1*, but are written as 'don't cares' – X. The *faulty* circuit will give a *0* output if both inputs to the OR are *0*. For the first

Fig. 4.8. Example of D-cubes of failure: (*a*) fault free; and (*b*) faulty.

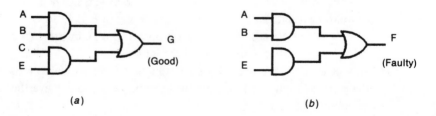

(*a*)   (*b*)

Table 4.2. *Fault free and faulty conditions in Fig. 4.1.*

A	B	C	E	G	F	A	B	C	E	G	F
1	1	X	X	1		0	X	0	X	0	
X	X	1	1	1		0	X	X	0	0	
						X	0	0	X	0	
0	X	X	0		0	X	0	X	0	0	
X	0	X	0		0						
						1	1	X	X		1
						X	X	X	1		1

(a)	(b)

one either A OR B must be *0* – it is immaterial which. For the second input to the OR, C or E must be *0* – C does not exist in the faulty function. If the C is thought of as s-a-*1* then what one *tries* to set it to is immaterial. Hence it is written as *X*. Table 4.2(*a*) represents the case of a fault free circuit with an output of *1* and a faulty circuit with an output of *0*. If a set of compatible inputs can be found, the output will be written D.

Table 4.2(*b*) is derived in a similar manner. A set of compatible inputs will result in an output of $\bar{D}$ – i.e. *0* in the fault free circuit and *1* in the faulty circuit.

Now attempt to **intersect** the two groups of vertices. This is done as follows. Start with the first row of Table 4.2(*a*) and compare (intersect) with the third row. If any pair of corresponding inputs are *1* and *0* then these two set of inputs are not compatible. For these two rows A for G is *1* and for F is *0*. The procedure is repeated for rows 1 and 4. Indeed, every row in the 'G' part of the table is intersected with every row in the 'F' part. For rows 1 and 4, B is *1* and *0* respectively, so the intersection is empty.

Now work with rows 2 and 3 and with 2 and 4. Here E is *1* in row 2 and *0* in both rows 3 and 4. The resulting set is still empty.

Moving to Table 4.2(*b*), intersect row 1 with row 5 – A is *0* and *1* respectively. Comparing rows 1 and 6, A and C both intersect as *0*, *X*. If *X* took the value *0*, as it may, the two are compatible, so write *0* in these columns. For B, (*X*, *X*) is recorded as *X* and is also compatible. E is *X*, *1*, so choose *X* to be *1* and write that. The complete resulting D-cube is

$$0 \ X \ 0 \ 1 \quad \bar{D}$$

$\bar{D}$ is chosen since G (for the good circuit) has *0* output and F (the faulty circuit) has *1* output.

Performing the full set of intersections on Table 4.2(*b*), rows 3 and 6 will give

$$X \ 0 \ 0 \ 1 \quad \bar{D}$$

and all other intersections are null. There is no intersection with a *D* output since all intersection in Table 4.2(*a*) failed.

The interpretation of the line *0 X 0 1* $\bar{D}$ is that, for a fault free circuit with inputs A = 0, C = 0 and E = 1, the output (G) is *0*, but that with these same inputs and a faulty circuit equivalent to that shown in Fig. 4.8(*b*), the output (F) is *1*. This is because the C input does not exist in the faulty circuit. The C value is not an *X* because in the fault free circuit C = 1 would lead to an output of *1*, not *0*. The signal set *X 0 0 1* $\bar{D}$ can be interpreted in a similar way.

The conclusion to be drawn is that, if either of A or B is set to *0*, E is set to *1* and we *try* to set C to *0*, then the fault free and faulty circuits will have different outputs. Hence the fault can be detected.

Notice once again that the cause of the fault is not part of this discussion. Any fault resulting in the function F will be tested by either or both of the signal patterns derived.

To fix ideas a little more, consider the two-input AND gate of Fig. 4.9(*a*) and the faulty gates of Figs. 4.9(*b*) and (*c*). In Fig. 4.9(*b*) A is stuck at *1*. Table 4.3(*a*) shows the cubes.

Take the first case of a s-a-*1* on A, Table 4.3(*a*). In the *fault free* case G is *1* if both inputs are *1* and *0* if either input is *0*. For the *faulty* gate, the output is the same as B regardless of A. For *D* there is no successful intersection. For $\bar{D}$, rows 1 and 3 lead to

$$0 \ 1 \quad \bar{D}$$

Thus, if A is *0* and B is *1*, the fault free circuit gives a *0* out, whilst the faulty circuit gives a *1* since A is s-a-*1*.

For the case of A s-a-*0*, Table 4.3(*b*), the faulty output is always *0*. Thus the faulty circuit never has a *1* output, and the set of signals for $\bar{D}$ is empty.

Fig. 4.9. Two-input AND gate (*a*) with one input s-a-*1* (*b*) and with one input s-a-*0* (*c*).

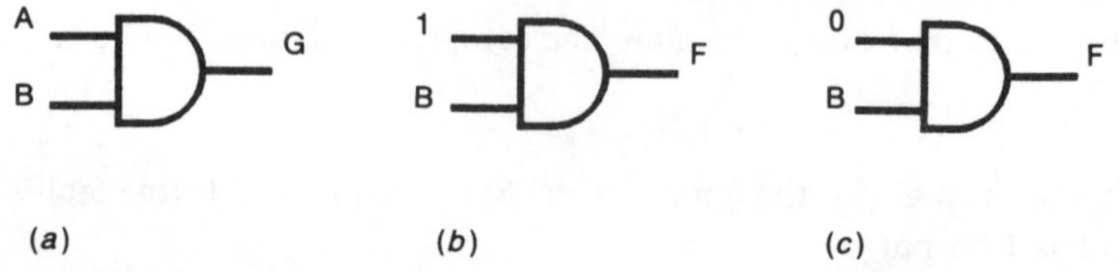

(*a*)   (*b*)   (*c*)

Table 4.3. *Cubes for a two-input AND gate:* (a) s-a-*1* and (b) s-a-*0*

	A	B	G	F		A	B	G	F
For *D*	*1*	*1*	*1*		For *D*	*1*	*1*	*1*	
	X	*0*		*0*		*X*	*X*		*0*
For *D̄*	*0*	*X*	*0*		For *D̄*	*0*	*X*	*0*	
	X	*0*	*0*			*X*	*0*	*0*	
	X	*1*		*1*		empty			

(a)   (b)

The set of inputs for *D*, however, intersects to give the vertex

> *1 1   D*

In other words, if A and B are both *1* the good circuit has a *1* output whilst the faulty circuit with A s-a-*0* has a *0*.

4.4.3 Primitive D-cubes of a logic block (**propagation D-cubes**)

In the previous section a method of finding a test for a fault in a logic block was specified. This resulted in a *D* or *D̄* on the output. It is now necessary to consider how the inputs derived above can be traced back to derive the primary inputs to the network (point (b) in Section 4.4.1) and how the fault may be propagated to a primary output of the network as indicated in point (c) of Section 4.4.1.

Table 4.4 gives a full set of cubes for the two-input AND gate. Table 4.4(*a*) shows the usual truth table giving the primitive cubes. Tables 4.4(*a*) and (*b*) are those shown in Table 4.1. *D D D* has been added. Clearly, if both inputs are *1* the output is *1* and if both inputs are *0*, so is the output. Table 4.4(c) shows the D-cubes of failure for A s-a-*0* and s-a-*1*. The cube *1 0 D̄* is for input B what *0 1 D̄* is for input A and hence detects B s-a-*1*. It will now be realised that the first two lines of Table 4.4(*c*) also detect G s-a-*1* and this cannot be distinguished from A or B s-a-*1*. Line 3 gives a test for the single fault G s-a-*1*, but cannot be distinguished from the double fault A s-a-*1* AND B s-a-*1*. Similarly, it is not possible to distinguish between one or both the inputs s-a-*0* or the output s-a-*0*.

Suppose input A of Fig. 4.8 was a *1* and was supplied from an AND gate. Table 4.4(*a*) shows that, for a *1* on the output, the inputs must all be *1*s (first line). Each of these inputs must then be further traced back. Conversely, if A

Table 4.4. *Cubes for a two-input AND gate:* (a) *primary cubes;* (b) *primitive D-cubes of logic;* (c) *primitive D-cubes of failure*

(a)

A	B	G
1	1	1
0	X	0
X	0	0

(b)

A	B	G
D	1	D
1	D	D
D	D	D

(c)

A	B	G	
0	1	$\bar{D}$	A s-a-*1* or G s-a-*1*
1	0	$\bar{D}$	B s-a-*1* or G s-a-*1*
0	0	$\bar{D}$	(A & B) or G s-a-*1*
1	1	D	A or B or G s-a-*0*

had been specified as a *0*, then lines 2 and 3 of Table 4.4(*a*) show that a *0* on either input is required and the other input is a 'don't care' – *X*. Only the input specified as *0* need be followed further.

To trace forward to the primary outputs, Table 4.4(*b*) is used. Suppose that the output in Fig. 4.8 is specified to be *D* and the following gate is an AND gate. Table 4.4(*b*) shows that, to propagate *D* to the output, the other input(s) must be set to *1*s. This implies that those inputs must be traced back to the primary inputs as described in the previous paragraph.

To fix ideas in relation to sets of cubes, the reader might like to derive the cubes for a three-input OR gate. Table 4.5 gives the result.

Table 4.5. *Cubes of a three-input OR gate*: (a) *primary cubes*; (b) *primitive D-cubes of logic*; (c) *primitive D-cubes of failure*

(a)

A	B	C	G
0	0	0	0
1	X	X	1
X	1	X	1
X	X	1	1

(b)

A	B	C	G
D	0	0	D
0	D	0	D
0	0	D	D
D	D	0	D
D	0	D	D
0	D	D	D
D	D	D	D

(c)

A	B	C	G	
1	1	1	D	(A & B & C) or G s-a-0
1	0	0	D	A s-a-0 or G s-a-0
0	1	0	D	B s-a-0 or G s-a-0
0	0	1	D	C s-a-0 or G s-a-0
1	1	0	D	A & B s-a-0 or G s-a-0
1	0	1	D	A & C s-a-0 or G s-a-0
0	1	1	D	B & C s-a-0 or G s-a-0
0	0	0	$\bar{D}$	A or B or C or G s-a-1

4.4.4 Example of use of D-cubes

As an example, return to the circuit of Fig. 3.16, redrawn as Fig. 4.10. The procedure is similar to that of Section 3.6.1. Table 4.6 reproduces the cube tables for two-input AND and OR gates. We wish to test for W stuck at *1*.

First specify W to be $\bar{D}-1$ in the faulty circuit. Attempt to set it to *0* for a fault free circuit.

Now try to propagate this to the output. To do so we use the propagation D-cube of an AND gate.

The first input is $\bar{D}$. For this purpose D and $\bar{D}$ are interchangeable. The first propagation $\bar{D}$-cube in which the second input is *1* is required, giving the cube $\bar{D}$ *1* $\bar{D}$. As the primary output has been reached, this phase of the test generation is complete. It is called the **D-drive** phase and is shown in the second line of Table 4.7. In this table all signals are initially set to X, and W is set to $\bar{D}$.

Now attempt to **justify** all the other required signals. Starting from the output, the first (and only) signal to be justified is N. What is required is to select a set of primary inputs such that N is a *1* as required by the D-drive process. N is driven by an OR gate. The primitive cubes of the OR gate show that a *1* on either input will justify the *1* on the output (second and third line of Table 4.6(*b*)). For the sake of example choose L, Table 4.7, line 3. An automatic generator will choose one of the two possibilities at random, presumably the first that it finds in its data base (but see below).

L is the output of an AND gate. Observe that the primitive cubes of the AND gate require both E and F to be *1* to justify the *1* at L (line 3 of

Fig. 4.10. Example illustrating the use of D-cubes.

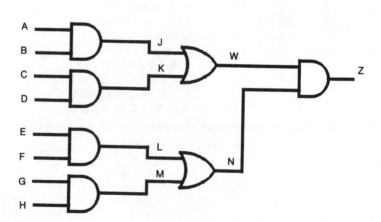

Table 4.6. *Cubes for two-input AND (a) and OR (b) gates*

(a)

Primitive cubes

0	X	0
X	0	0
1	1	1

Primitive D-cubes of logic

D	1	D	i/p or o/p s-a-*0*
1	D	D	i/p or o/p s-a-*0*
D	D	D	

Primitive D-cubes of failure

0	1	$\bar{D}$	A or G s-a-*1*
1	0	$\bar{D}$	B or G s-a-*1*
0	0	$\bar{D}$	(A & B) or G s-a-*1*
1	1	D	A or B or G s-a-*0*

(b)

Primitive cubes

0	0	0
X	1	1
1	X	1

Primitive D-cubes of logic

D	0	D
0	D	D
D	D	D

Primitive D-cubes of failure

1	0	D	A or G s-a-*0*
0	1	D	B or G s-a-*0*
1	1	D	(A & B) or G s-a-*0*
0	0	$\bar{D}$	A or B or G s-a-*1*

Table 4.7. *Signal generation for Fig. 4.9*

A	B	C	D	E	F	G	H	J	K	L	M	N	W	Z	Comment
X	*X*	*X*	*X*	*X*	*X*	*X*	*X*	*X*	*X*	*X*	*X*	*X*	$\bar{D}$	*X*	initialise
												1	$\bar{D}$	$\bar{D}$	D-drive
										1		*1*			justify N
			1	*1*						*1*					justify L
								0	*0*				$\bar{D}$		justify W = $\bar{D}$
0								*0*							justify J
		0							*0*						justify K

Table 4.6(*a*)). This is shown in Table 4.7 line 4. As E and F are primary inputs, this path is complete.

It is now necessary to return to W and to justify a *0* for the good circuit and *1* for the faulty circuit back to the primary inputs. The gate whose output is W is the faulty gate. The three steps which follow are shown in Table 4.7 lines 5 to 7. Observe the primitive D-cubes of failure of the OR gate, Table 4.6(*b*). To get $\bar{D}$ on the output we require *0* on *both* inputs. Begin by processing J, so first mark K for later processing. J must be a *0*. Observing the primitive cubes of the AND gate making J, it is seen that A or B must be *0*. Choose A – quite arbitrarily. This is a primary input, so the path is justified.

Now return to find any markers. There is one, namely that at K. This also needs to be *0*, so C is chosen to be *0*. All paths are now justified since a search will find no more markers. Hence a test for W stuck at *1* is

A B C D E F G H
0 *X* 0 *X* 1 1 *X* *X*

In practice, problems arise with reconvergent fan out. This occurs when one signal fans out to two or more places and later in the network some of the resultant signals combine. An exclusive OR circuit built from AND and OR gates always contains such problems (some EXOR ICs have special circuits which avoid these).

Consider Fig. 4.11. Generate a test for S stuck at *1*. For the fault free circuit S must be set to *0*, so place a $\bar{D}$ on that signal. The D-drive will put *D* on U. Justifying this requires *1* on T and hence *0* on R or Q. Let us choose to place the *0* on R with Q as a 'don't care' (*X*), since R is the first input of the gate producing T. This is shown in line 2 of Table 4.8. Justifying R requires *1* on P and Q, Table 4.8 line 3. Now return to S. To get $\bar{D}$ on S, that is, *0* in the good circuit, a *1* is needed on both P and R; but a *0* has already been

Table 4.8. *Signal generation for Fig. 4.11*

P	Q	R	S	T	U	
X	X	X	$\bar{D}$	1	D	D-drive
	X	0		1		mark as choice point
1	1	0				
1		1	$\bar{D}$			Not consistent; return to choice point
—	—	—	—	—	—	for alternative
	0	X		1		
1		1	$\bar{D}$			
	0	1				

assigned to R in line 3. Hence lines 3 and 4 of Table 4.8 are inconsistent.

When this happens, it is necessary to return to the last place where a choice was made. In this case, that place was in justifying T. The choice of R or Q to be *0* fell to R. That point in the process should also have been marked so that the return now required can be made. At this point, make the alternative choice. Q is set to *0* and R is set to *X* as shown below the line in Table 4.8. Q is a primary input, so this phase of the test generation ends.

Return to making another attempt to justify S. This requires *1* on P and R which is now fully consistent with the previous decisions. R set to *1* requires P or Q to be *0*. P and Q are already *1* and *0* respectively as required. The test for a s-a-*1* at S is

$$P = 1, \qquad Q = 0$$

4.4.5 Enhancements to the D-algorithm

The D-algorithm was first published in 1966. Since that time there have been many papers attempting to improve on the methods of

Fig. 4.11. Exclusive OR gate made with NAND gates.

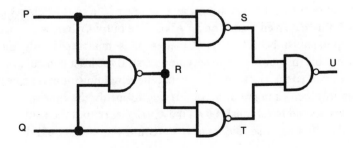

generating these patterns. So far as this author can see, the vast majority use the same *principle*. The differences lie mainly in the way that the D-drive and calculation of primary inputs is carried out. Goel (1981) discovered that the original approach was very slow for a particular class of circuits, namely, those using exclusive ORs, and derived an alternative in a scheme known as PODEM.

In PODEM one starts, as before, with the fault. Attempt to set the input of the source gate to the required value, and trace that back to a primary input. If this cannot be done, time is not wasted doing the D-drive. Having reached the primary input a 'simulation' of the circuit is performed, establishing all other signals in the network affected by this primary input. Doing this may set or block other paths that will be required later, so some incompatible choices can be avoided.

This procedure is repeated for all other signals from the fault until the paths from the primary inputs to the fault are established. Thus the element inputs for the faulty element are controllable by the primary inputs.

Now cause the fault D or $\bar{D}$ to move towards the *nearest* primary output by one logical element. This is called moving the **D-frontier**. Any signals required to establish that path are then justified. Again, the effect of all primary inputs set is simulated and checks made for consistency with previous settings.

Returning to Fig. 4.11, with S s-a-*1* we require P and R to be *1*. P is a primary input. As Q is not specified, no further simulation can be done from P. For R to be *1*, either P or Q must be *0*. P is already *1* so Q must be set to *0*. Simulating forward from here causes T to be set to *1*. This is what is required for the D-drive as the D-frontier is moved to U. There are no false choices and no backtracking.

The process continues until the D-drive is complete. The advantage here is that, if the inputs to the faulty element cannot be controlled, the D-drive never happens. Furthermore, if, after the D-frontier is moved forward, signals cannot be justified, then the D-drive is stopped. Thus it is probable that the number of false choices and hence amount of wasted work are reduced.

In performing the calculations, an estimate of the controllability of each signal is used. Given a fault on an AND gate output where a *1* is required, all inputs must be set to *1*. An attempt is made to justify the least controllable (most difficult to set) input first. If the least controllable input can be justified then it is quite likely that the remaining inputs can be also. If it cannot, then time is not wasted trying to justify the others.

If it is desired to establish a *0* on the AND gate output then any input at *0* will do. In this case the easiest to control input should be chosen.

Corresponding choices for an OR gate are to choose the most controllable input when establishing a *1* and the least controllable input first when establishing a *0*.

The process can be further improved. Consider Fig. 4.12 and suppose the test is for L s-a-*1*. For this to be tested it is necessary to specify L as $\bar{D}$, and hence H, K and E are all *1*. PODEM may select to trace back K first as the most difficult. This leads to J=0 and hence (say) B=0. B=0 implies H=0, which is inconsistent with the requirement that H be *1*. PODEM then has to start again and choose C to be *0* and B and A to be *1*. In FAN (Fujiwara and Shimono 1983) all signals forming L are traced back together. Thus

    Step 1:   H=K=E=*1*;
    Step 2:   J=0; A=B=*1*:
    Step 3:   C=0.

A second improvement in FAN is to notice situations such as that illustrated in Fig. 4.13. Here, elements E1 and E2 are on the path to the primary output regardless of the splitting and reconverging of the path. Hence these two elements are processed before either of the two split paths,

Fig. 4.12. Example for modification of PODEM as in FAN.

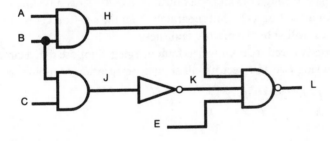

Fig. 4.13. Two paths to same primary output.

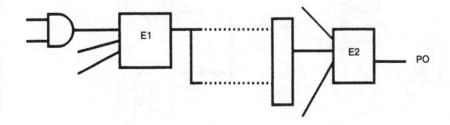

since if either is impossible to justify there is no point in following the complexities of the multiple paths.

4.5   Reducing the number of test vectors

The D-algorithm and most of its derivatives will find a test for a given fault if such a test exists. The type of fault is not specified, but the faulty function that results from the fault must be known. The network under test is restricted to combinational logic. There must be no feedback within the section of logic concerned. It has already been suggested that good design practice will confine feedback to sections of logic separated by registers, or provide means by which such feedback paths can be broken during testing.

The number of tests that must be performed can be reduced in a number of ways. For example, s-a-0 on the input of a NAND gate cannot be distinguished from a s-a-1 on the output or from a s-a-0 on any one or more other inputs. Thus there is no point in testing for the s-a-0 faults on the input. This is known as **fault collapsing**. Of course, these signals are the outputs of other circuit elements. A test for s-a faults on these outputs is required. Consider Fig. 4.14. For a TTL circuit, the second input of G2 will appear to be s-a-1. However, it is possible that a test for G1 output s-a-1 will use the path through G3 and will not show a fault. Thus it is still necessary to test for a s-a-1 on G2 input separately from the s-a-1 on G1 output. The distinction will show where the fault lies.

A second procedure is known as **fault merging**. Suppose that a circuit has the following two test vectors, all signals being primary inputs.

$$1 \quad 0 \quad X \quad X \quad 0 \quad X \quad 1$$
$$X \quad 0 \quad 1 \quad X \quad X \quad 0 \quad 1$$

Fig. 4.14. s-a-1 on G2 input not found by s-a-1 on G1 output.

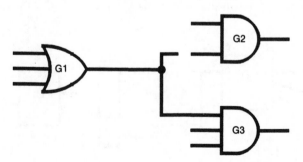

It is seen that where a defined signal appears in both vectors it is the same – both *0* or both *1*. All other signals have an *X* in at least one of the vectors. Hence these two vectors can be merged into a single vector which tests for all faults found by the two vectors separately. This will be

$$1 \quad 0 \quad 1 \quad X \quad 0 \quad 0 \quad 1$$

Where one of the corresponding signals is a known value (*0* or *1*) and one is *X*, the known value is used. If both vectors have an *X* then an *X* is still appropriate. An example has been quoted of the SN74LS630 (McClusky 1985). This has 23 inputs giving about eight million tests. However, each of the six outputs depends on no more than ten inputs. Hence 6 * 1K tests is sufficient. It is also possible to do each of the six tests in parallel, so only 1024 patterns are required.

It was suggested earlier that test vectors generated manually or using random patterns could rapidly build up 70–90% fault cover with relatively small numbers of vectors. The best approach to test generation is to use one of these methods and then to assess the fault cover (see Chapter 9). In the case of randomly generated tests, this may be done after every so many vectors have been generated until a desired fault cover has been obtained or until the improvement over the last assessment is 'small'. The fault cover assessment will provide a list of faults that are found by the test set – a **fault dictionary**, or better, a list of faults which have *not* been tested. This list can then be used as input to an automatic test generation program which only needs to find tests for the faults still untested. The automatic program should begin with fault collapsing to eliminate equivalent faults which cannot be separated, and the complete set of tests, including the manual/random pattern generated tests should be subject to fault merging.

A further suggestion is that the automatic test program generator (ATPG) should start with faults near the primary inputs. This then has long sensitised paths to the primary outputs, and hence many other possible faults on this path can be included with a minimum of computation – only the setting of D or $\bar{D}$ on the path needs to be varied.

The program CONT (Takamatsu and Kinoshita 1989) adopts another approach. It begins by trying to find primary inputs as in PODEM. However, when it finds it cannot make progress, it does not backtrack to the original fault. Instead it tries to discover another fault which the current set of primary inputs can detect. In this process a D will be replaced by a *1* and a $\bar{D}$ by a *0* in appropriate places. The reference gives a good example.

4.6 The MOS stuck open fault

So far it has been assumed that faults can be modelled as a stuck at *1* or stuck at *0*. There is another fault in MOS circuits which cannot be modelled in this way (Elziq 1981). Consider the CMOS circuit in Fig. 4.15(*a*) and suppose that there is a broken wire at A. Fig. 4.15(*b*) shows the equivalent gate symbol. Let us try to find a s-a fault which will find the **stuck open** fault.

- Test for Z s-a-*1*. This requires an attempt to set Z *low*, requiring X or Y to be *high*. If Y goes *high*, Z will go *low* as for a good circuit. If X now goes *high*, Z will stay in its previous state, whatever that was, since T4 is turned off. This is unsatisfactory as a test for a stuck open, but suggests that a test for a fault on X might be more useful.

- Test for Z s-a-*0*. An attempt is made to pull Z *high*, so both inputs must be *low*. The circuit with A stuck open behaves as the good circuit, so that stuck open is clearly not detectable.

- Test for X s-a-*1*. Y is set to *low* to allow X to be the controlling input. Try to set X to *low*. Z goes *high* in both the fault free circuit and that with A stuck open, since both T3 and T4 will be on. If X were stuck *high* in an otherwise good circuit then Z would stay *low*. This will not detect the stuck open.

- Test for X s-a-*0*. Y is still *low* and try to set X *high*. In the good circuit Z goes *low* and if X is s-a-0 Z becomes *high*. With A stuck open Z retains its previous value. Again, this is not a satisfactory test in itself.

Fig. 4.15. CMOS NOR gate.

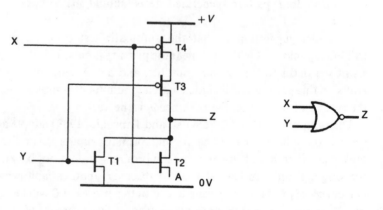

(*a*)

(*b*)

Thus a test for a s-a fault will not find the fault caused by the break at A in Fig. 4.15(a). The reason is that the fault has caused a change of logic function to

$$Z = \bar{X}\bar{Y} + X\bar{Y}Z_n$$

where Z_n is the previous value of Z. The circuit has become a sequential circuit as shown in Fig. 4.16. A test for a s-a fault on one of the wires to the OR gate will find the stuck open fault, but these are not 'real' gates. To use this model in simulation of MOS systems makes them very complicated to simulate.

The reason for this problem is that an MOS gate input requires very low current indeed to drive it. Consequently output nodes such as Z drive purely capacitive loads, and if none of the three paths to a power supply conducts, Z remains at its previous value. This is in contrast to a TTL gate, for example, where an open circuit output transistor drives inputs which take a significant current, and which would pull the gate output high (Fig. 4.17). Thus a test for Z s-a-1 is sufficient.

Fig. 4.16. Model of CMOS NOR gate.

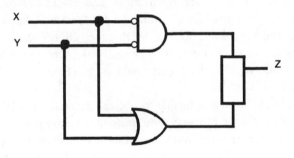

Fig. 4.17. TTL equivalent to stuck-open circuit.

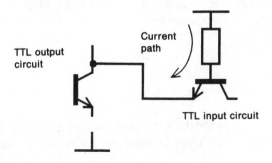

It is worth considering briefly other possible faults. A stuck open in T3 or T4 of Fig. 4.15 would be found by a s-a-*0* test on Z provided that X or Y had at some point been taken *high* first, pulling Z *low*. The storage on Z would ensure it never went *high* again. This requires at least one extra test vector at some point and hence probably a pair of vectors per fault. Stuck closed faults would alter the logic function, assuming that, if T3, T4 and one of T1 and T2 were conducting, the result on Z was either a *1* or a *0* and not an intermediate value. If the latter occurs then the circuit is untestable by digital means. The D-algorithm can handle changes of logic function, but it is not simply a matter of a s-a fault.

The points being made here are:

- the s-a model of faults at gate level is not always sufficient,
- there may be other types of fault in all types of logic which could be untestable with the algorithms traditionally described.

Experience to date suggests that the s-a model works in most cases, including many MOS faults. For MOS circuits, some additional tests are required to cope with stuck open faults. Elziq used a search for pairs of vectors needed to find the stuck open faults, and where they were not present in the tests generated for s-a faults, he put them in. Gheewala has discovered 26 different faults for an MOS gate, all of which can be represented by the set of fault models suggested here. On the other hand, Shen and Hirschhorn (1987) have cast doubt on whether the stuck open fault deserves attention.

Unfortunately there is one further problem. Because an MOS gate output possesses a storage property, any transient switching of the circuit could 'store' to the wrong state, much as a random asynchronous input to a flip-flop (see Fig. 4.16). This implies that great care must be taken to ensure that inputs change in the correct order. Since most test machines (as opposed to simulators) cannot guarantee the order of changes on a fine time scale, it may be necessary to specify intermediate patterns. For example, suppose it is desired to change X Y from *0 0* to *1 1* but at all costs avoiding the pattern *0 1*. The sequence

0 0
1 0
1 1

must be specified. For some testers, forcing this means forcing an extra test with the intermediate pattern. There are now *three* test vectors for one fault!

4.7 High level testing

As systems – or even ICs – become larger, the number of possible transistor or gate faults gets out of hand, even accepting the possibility of partitioning the logic into more testable blocks. There is a clear need to extend testing methods to high level blocks of logic.

The second reason for high level testing is to reduce the time taken when fault finding in the field. The priority is to keep expensive equipment running, by keeping the test equipment simple. Once the faulty unit is found it is 'repaired' by replacement, and if repair is done at all, it can be done in a workshop with full facilities. There is only one of these.

There is a third reason for using high level testing. With large modules, such as microprocessors, the fault modes and logic models are unknown. Hence it is impossible to apply gate level test procedures.

Where the total logical system is being designed, an approach is to divide the system into easily testable blocks (Noujain 1984). It is presumed that these blocks would be smaller than the combinational blocks of a scan designed system. One might think in terms of 4-bit adders. A set of test vectors for each block capable of achieving 100% fault coverage is designed. Groups of these blocks are then formed as shown in Fig. 4.18. Means are devised for setting the necessary patterns at the inputs of block A by setting the primary inputs to blocks B, C and D to get the appropriate outputs. Inputs to B and D must also set the inputs to blocks P and Q in such a way as to propagate the outputs of block A to the primary outputs of the total network.

This second level block can now be combined with others in a hierarchical system.

Chandra and Patel (1987) produced HIPODEM, a version of PODEM,

Fig. 4.18. Hierarchical test of block A.

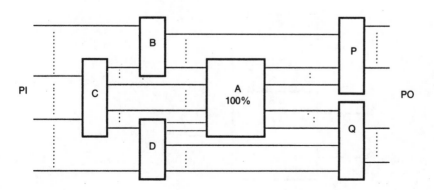

to operate in this manner. A set of D-cubes for the adders etc. is produced, as well as controllability and observability measures.

Consider, a combinational module such as that in Fig. 4.19 (Murray and Hayes 1990). Firstly, it is necessary to know the value of the controls to propagate signals from data (in) to Z (out). Suppose a set of data, D, results in an output, Z. Attempt to find a control input such that for any set of data D′ different to D the value of Z′ will be different to Z. For example, in the multiplexer shown in Fig. 4.20, if $C=0$, $Z=X_1$, so X_1, Z is a 'pair.' If $C=1$, $Z=X_2$. Furthermore, if $X_1=0$ and $X_2=1$ then C, Z is a unique pair. This circuit can be tested, therefore, and a set of D-cubes can be constructed as shown in Table 4.9.

Unfortunately many of the higher level blocks that might be considered are not purely combinational. Shift registers and counters, not to mention processors, come immediately to mind. These not only have inputs and outputs but also input *states*, the latter two of which are dependent on sequences of inputs as well as the history of internal states and outputs. Breuer and Freedman (1980) suggest an algorithmic solution for determin-

Fig. 4.19. Testing a combinational logic block.

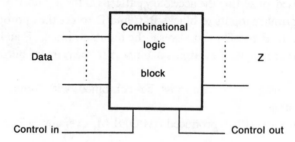

Fig. 4.20. A multiplexer.

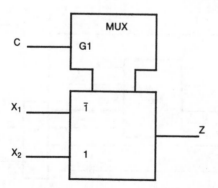

Table 4.9. *Cubes for the multiplexer of Fig. 4.20*

X_1	X_2	C	Z
0	X	0	0
1	X	0	1
X	0	1	0
X	1	1	1
D	X	0	D
X	D	1	D
0	1	D	D

ing the operation. They claim that the algorithms are often applicable to a whole class of problems. They develop a language for expressing both sequential and parallel events. D-drive and line justification are all time dependent and may run into inconsistency problems in time as well as in 'space.' The reader is referred to the paper for a worked example.

Since 1990 the number of papers in this area has expanded rapidly. Ghosh *et al.* (1991) and Sarfert *et al.* (1992) have both produced interesting work. This expansion is in recognition of the fact that gate or lower level testing of very large logic systems, including chips, is becoming progressively more expensive.

This chapter has given an introduction to test program generation. It may seem strange that a main technique described is 25 years old. The reason is that no fundamentally different method has been discovered. An understanding of what has been described is sufficient to enable all newer proposals to be understood quite easily.

5

Input/output of simulation and specification of models

5.1   Input and output of simulation

When a complete system is being simulated there are relatively few external inputs – a start key, a break key and possibly some inputs from various peripherals such as tape drives, sensors etc. Similarly, there are few outputs. In these cases the arrangements to pass 'test data' to the simulation can be very crude.

The main purpose of simulation is to find errors in a design. To simulate a complete system and expect to find detailed errors is very difficult, especially as different parts of the design may be at different stages of development. It is important to be able to simulate sub-units independently of the total design in order to get the majority of problems solved before trying to integrate the complete system (Fig. 1.4). Some means of supplying and controlling test vectors is required.

In this form of testing there will be copious output, which comes in two forms. The first is the values of primary output signals – the product of a multiplier, for example. These values can be compared with a set of 'expected' values to check the overall operation. If the results of simulation are different to the expected values then it is necessary to

- check the expected values – *Note*!!!
- if the expected values seem correct, trace back through the logic to find where the error occurred and hence find the design error.

Therefore, secondly, internal signals must be available. This is always possible with a simulator whereas it may be impossible in testing hardware, especially ICs. Thus a simulator requires facilities for monitoring all signals and for displaying them in a form which enables the user to trace through the network easily.

This emphasises two features of a simulator which are not available with prototype hardware.

- Internal logic states can be monitored.
- Sub-units of the logic can be isolated and tested on their own.

In this chapter, an indication of the control of a simulation is given. The description will make use of the ideas and notation of **VHDL** (VHDL 1987, Perry 1991) – VHSIC[1] Hardware Description Language. It is not intended that this should be a definitive description of VHDL – it is not – and the reader must consult specialist texts for that. Having said that, the reader unfamiliar with VHDL should not despair. Many of the features will be clear to anyone familiar with a programming language. With each example there will be a suitable description.

VHDL is intended for describing hardware. Here the interest is solely in controlling and monitoring certain signals. Details of the structure of the hardware descriptions will be omitted. VHDL reserved words will be printed in upper case.

The first feature of note is that VHDL does *not* contain a mechanism for setting primary inputs to given values or for monitoring primary outputs. In other words, a complete system with no inputs or outputs is presumed. For a real design or a partial design the logic of interest is set alongside a piece of pseudo-logic which generates the primary inputs, Fig. 5.1. A second

Fig. 5.1. Simulation of a system described in VHDL.

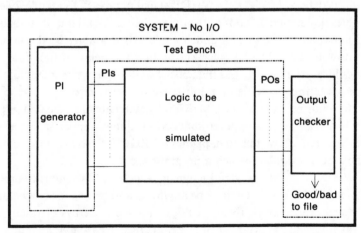

[1] Very High Scale Integrated Circuits – a project of the Department of Defense in the USA.

piece of pseudo-logic can be set up to check the outputs. These two are sometimes described as a **test bench**. Hence VHDL is a *modelling* language and not a driving instruction language.

As a rule, the simulation environment has facilities to monitor and display signals and variables within a hardware description and possibly to set primary inputs. This is not part of VHDL itself. VHDL does have a facility for text output and input but it is rather awkward in use. It is useful for good/bad indicators. There is also a mechanism for displaying messages (ASSERT).

The primary interest in the present chapter is in modelling the primary input (PI) generator (Fig. 5.1), but it will be useful to indicate facilities for models of a more general nature.

5.2   Simple driver

In its crudest form, the primary input generator consists of a series of statements setting each of the signals to a value. There is then a statement requiring no further changes until the logic being simulated has had a chance to settle to its new state. Thus one has statements such as

```
a <. = '0';
b < = '1';
WAIT FOR 50 NS:
```

In this series of statements a and b are two input signals. They are of type BIT as defined in VHDL. A VHDL signal of type BIT can take one of two values designated '0' and '1,' the quotes being essential to distinguish from the characters 0 and 1. a and b could, for example, be the inputs to a two-input gate. The designer knows that the delay of the piece of logic should be significantly less than 50 ns, so the WAIT statement allows time for the output to settle even in the face of some error.

In a normal ARCHITECTURE all statements are executed in parallel. In this example it is required that certain things happen in sequence. The feature that allows this to happen is a PROCESS, so the statements above must be embedded in such a program segment.

VHDL is a very powerful language which allows signals to have many different TYPES. These may be anything, and can be (and often are) user defined. They can be INTEGERS, colours, etc. Arrays of values are also available, and in particular BIT_VECTORS. This allows a set of signals of type BIT to be used together. For example, one could define

```
SUBTYPE two_bit IS BIT_VECTOR (0 TO 1);
```

and then have a signal definition

SIGNAL D : two_bit;

whence the setting of the two input signals of the above example can be written

D < = B"10";

The B here implies the following quoted string is a set of values in binary. The string must have a length which exactly matches the type definition – there is no assumption of left extension with zeros. However, there is a general mechanism by which a short vector can be inserted in or extracted from a longer one.

A question arises as to what causes a simulation to stop once started. If allowed to run on 'for ever' much CPU resource (and hence money) is wasted. In the very simple example just quoted, it is clear that all activity in the circuit ceases at 50 ns, and this circumstance might be detected. However, suppose there to be an error – possibly a typing error – causing the output to be connected back to an input such as to cause an oscillation. Looking for no activity in the circuit would never stop the simulation. Similarly if two clock signals were used for testing the circuit. There is only one safe procedure. *The environment of the simulation must have a stopping mechanism.*

For example, on one system the 'simulation control' asks for how long the simulation is to be run. The default is 0 ns. The user needs to be judicious in choosing this value to conserve resources. Too long a run wastes money. Too short a run will not achieve the objectives and a further run will be needed, which is equally wasteful. This stopping mechanism is essential. It is *not* part of VHDL.

To complete this example, suppose that the logic to be simulated in Fig. 5.1 is a two-input AND, and suppose this has been described in the appropriate manner as an ENTITY called 2_input_and. Prog. 5.1 shows how a test might be set up. There is an ARCHITECTURE called *syst* of an ENTITY called *system*. It consists of an instance, C1 of the component 2_input_and. The signals x, y and z in the COMPONENT definition are formal parameters, while a, b and c are the actual signals for the specific instance C1. To produce the device inputs, a PROCESS is used. It is given a name (optional), pi_gen. The inputs are assigned and there is a WAIT FOR the outputs to settle. The next statement, the ASSERT, does *nothing* if the condition c = '0' is true, which it should be. If the logic of 2_input_and is faulty, then a REPORT is produced. This is the output checker of Fig. 5.1. A second set of inputs is then asserted and a further WAIT FOR and test of the output made. This is written in a different form to show a facility of VHDL, but the effect is the same. Both tests could be written this way. Alternatively the second test could be written

ASSERT (c = '1')....

The final WAIT is a means to allow the PROCESS to end, but it waits for ever. Thus the environment time limit on simulation *is essential*. The test can be easily extended to the other two input possibilities.

It is worth pointing out that this simple test is a test of a logic subsystem. Once the designer has decided it is acceptable 2_input_and can be incorporated as a COMPONENT in other ARCHITECTURES, knowing that its internal operation is correct to some degree.

5.3   Simulation output

The output of a good simulation as described will be nothing. Such an output leaves the user with a degree of uncertainty. There could be a mistake in the monitor part of the program. One student project observed by the author actually had the monitoring turned off! To obtain something

Prog. 5.1. ARCHITECTURE of Fig. 5.1.

```
ARCHITECTURE syst OF system IS
COMPONENT 2_input_and
    PORT (x, y : IN BIT; z : OUT BIT);
END COMPONENT;

SIGNAL a, b, c : BIT;
BEGIN
    C1 : 2_input_and
        PORT MAP (a, b, c);

    pi_gen : PROCESS
    BEGIN
        a < = '0';
        b < = '1';
        WAIT FOR 50 NS;
        ASSERT (c = '0') REPORT "c = '1' from a = '1', b = '0' "
            SEVERITY ERROR;
        a < = '1';
        WAIT FOR 50 NS;
        ASSERT (c = (a AND b)) REPORT "c = '0' from a = b = '1'"
            SEVERITY ERROR;
        WAIT;
    END PROCESS;
END syst;
```

more tangible, and to be able to fault find, it is necessary to be able to observe some signals. On a small network, for a limited simulation, it will be possible to observe all signals all the time. However, consider a network of 1M signals simulated with a 1 ns time step for 1 ms. If each piece of monitor data is only one byte, the output is 10^{12} bytes. This is clearly nonsense. Even if it can be stored, it cannot be read by any real user in a finite time.

There are several ways of reducing the quantity of data.

- Only record changes. With system activity of 1% in an event driven simulator (Chapter 6) the number of data items is reduced by a factor of 100. Time must also be recorded with each set of data. This is not a big enough reduction.
- Only certain time periods are of interest. In an initial run record results only at 'clock' times. If an error is detected, only switch on recording for the period just prior to where the error occurs.
- Restrict the number of signals being monitored. In a hierarchical description, limit recording in the initial run to high level signals. When an error occurs, rerun looking at more detailed levels of the design in the area where the error happened.

Suppose a large system as above is being simulated. Suppose it contains an arithmetic unit with an adder and a multiplier (e.g. Fig. 1.4). The arithmetic unit output is one of several sets of signals monitored. Let the monitor interval be 500 ns. At 25 µs the unit outputs are all as expected but at 25.5 µs the arithmetic unit outputs show a difference. It may be possible to see that the arithmetic unit was doing a multiplication rather than an addition. Assuming the fault cannot be further identified immediately, the simulation should now be rerun without monitoring up to 24.5 µs, say. Monitoring of the arithmetic unit outputs, and also of the multiplier outputs and some of the multiplier internal signals, is then turned on. If it is still not possible to monitor all of these, a judicious choice will help. For example, if the error begins in the less significant part of the result then it is in this area that the most detailed monitoring should be done. The simulation is continued only as far as 25.5 µs. Alternatively it might be possible to simulate the multiplier on its own with the operands from the original simulation. This will save even more resources.

VHDL has very limited facilities for output. The ASSERT statement has been mentioned. There is also provision to write to a file. This is not entirely satisfactory. In practice implementors of VHDL are providing environments which enable signals or variables to be marked for monitoring and then to display these in either tabular or graphical form or both. The ability to look at a short period of time selected from a longer one is available (e.g. 24.5 μs to 25.5 μs from 1 ms).

The signals to be monitored must inevitably be selected prior to simulation. Consequently the 'wrong' signals will be selected quite often, and the simulation will have to be rerun with a different selection. In a bad case it may take several runs to find the correct selection. In some environments it may be possible to set a **break point**. At the break point *every* signal in the network at all levels is recorded, together with any other data held within the simulator (see Chapter 6). In the example the break point would be at 24.5 μs. It is now possible to reload these values into the simulator and start the simulation from the break point, knowing that the simulator will hold correct values and simulation history at that point. Further, it will be possible to restart from the break point as often as may be necessary. Thus repeated running of bits of simulation which are known to be 'good' is avoided.

Returning to the example of Prog. 5.1, one might choose to monitor all signals for all times, since the example is very simple. The resulting output may appear as Table 5.1 or as Fig. 5.2. Table 5.1 is a textual form showing the times at which signal changes occur. t_{phl} is set to 23 units and t_{plh} to 15 units. The second form shown in Fig. 5.2 is a waveform diagram as might be observed on an oscilloscope or logic analyser.

5.4   Operation in parallel and in sequence

The simulator control commands so far introduced are probably sufficient to enable most things to be done. In particular, automatically generated test programs will generally produce large sets of test vectors which are applied to the primary inputs at appropriate intervals. Such test

Fig. 5.2. Graphical output of simulation of Prog. 5.1.

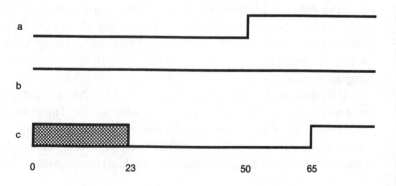

Table 5.1. *Tabular output of the simulation of Prog. 5.1*

time 0	$a=0$	$b=1$	$z=?$
time 23	$a=0$	$b=1$	$z=0$
time 50	$a=1$	$b=1$	$z=0$
time 65	$a=1$	$b=1$	$z=1$

programs are relatively 'unintelligent' and may not be very useful as test sets for debugging the design.

In design checking, applying test sets using only the simple commands of the previous section could be very cumbersome. Consider, for example, a free running clock of period 100 ns. This requires a set of instructions as shown in Prog. 5.2, interspersed among other drive waveforms and repeated every 100 ns. This is clearly unsatisfactory. Two things are needed.

- The ability to separate out the clock driver from other driven signals.
- The means to have several drive sections running in parallel and, at the same time, to be able to force a correct time sequence.

The VHDL provisions to enable these have already been indicated in outline. All signal allocations occur in parallel (i.e. at the same time) unless otherwise specified. COMPONENTS within an ARCHITECTURE are equally simulated in parallel. The COMPOMENT is simply a high(er) level description of something which has a separate architectural description elsewhere. At the lowest level such a description will be of the form shown in Prog. 5.2. Thus in Prog. 5.1 the AND gate will have an ARCHITECTURE of the form shown in Prog. 5.3. In Prog. 5.1 there are two items which are evaluated in parallel. These are the gate, C1, for which Prog. 5.3 can be specified as the low level description (see Prog. 5.4) and the PROCESS pi_gen.

To force things to happen in correct time sequence, the PROCESS is used. In a PROCESS each statement is executed in sequence. In principle there is a time delay between each statement. If no specific time is mentioned then VHDL invents a special sub-interval called a **delta**.

Prog. 5.2. Simple clock generator.

```
clock < = '0';
WAIT FOR 50 NS;
clock < = '1';
WAIT FOR 50 NS;
```

Considering Prog. 5.1, the assignment to a occurs at time T, say, and that to b must then occur at time $T + \delta t$. There is then a WAIT FOR 50 NS.

The AND gate of Prog. 5.3 evaluates as follows. 'and_gate' is simply a name for the PROCESS and is optional. The term (x, y) is called a sensitivity list. The PROCESS will be evaluated whenever any signal in the sensitivity list changes value. The construction '--' means that the rest of the line is a comment. p is a local VARIABLE. A VARIABLE is different to a signal in a very significant sense. The assignment p := x AND y happens *immediately*, with no delta involved. If p had been a SIGNAL instead, then the value of p would become a prediction for the future (see Section 6.3). The IF clause would then operate on the *old* value of p. For example, suppose x = '1' and y = '0' and the simulation has settled down. p is '0'. Now let y change to '1'. If p were a signal then it would be scheduled to become '1' at $T + \delta t$. The IF clause then operates with the value p = '0'. By specifying p to be a VARIABLE, p becomes '1' immediately and the IF clause operates as is required with p = '1'.

All PROCESSes must have a sensitivity list or a WAIT statement. The WAIT statement causes the PROCESS to suspend until the specified conditions are met (a length of time in the examples so far, but see below). When the conditions are met the PROCESS is reactivated as if a signal in a sensitivity list had changed. In Prog. 5.1 there is a WAIT without a condition. This suspends the PROCESS for ever. PROCESSes may contain IF clauses as indicated, and also CASE and LOOP clauses as will appear below.

An important difference between VHDL and a conventional programming language should be noted. In a programming language there is a strong concept of 'control flow.' A 'main program' is in overall control, but passes that control temporarily to a sub-program – procedure, function etc. There may be a whole hierarchy of such sub-programs but only one can be

Prog. 5.3. ARCHITECTURE of an AND gate.

```
ARCHITECTURE and_2 OF 2_input_and IS
BEGIN
        and_gate : PROCESS (x, y) --'and' is a reserved word
            VARIABLE p : BIT;
        BEGIN
            p := x AND y;
            IF (p = '1') THEN z < = p AFTER tplh;
                ELSE z < = p AFTER tphl;
            END IF;
        END PROCESS;
    END and _2;
```

operating at any one time. In a circuit all elements are operating at all times and this must be reflected in the hardware description language. In VHDL there is no master control. All sub-programs (ENTITIES, PROCESSes, FUNCTIONs, PROCEDUREs) can be active at the same time and without reference to any of the others. Interactions are controlled by the network connections and by the passage of simulated time.

Prog. 5.4 shows another way of testing the AND gate. A generalised clock generator is defined. The ENTITY clock_gen has three GENERIC parameters which specify the mark ('1'), and space ('0') of the clock waveform and a parameter, stop, which causes the waveform to end. The associated ARCHITECTURE has several features.

- NOW is a VHDL variable for current time.
- The signal 'clock' can be set more than once in the same statement. In this case it is set to '0' immediately (after one δt) and to '1' after the space period.
- The parameters of WAIT FOR and other similar clauses can be expressions – in this case the sum of mark and space.

The clock generator will be entered at time zero. A full cycle is executed. The current time, NOW, is then tested against stop. If NOW is less than stop a complete further cycle is executed, even if stop is NOW + 1. When NOW is greater than stop on this test, control passes to an infinite WAIT which ends the operation of this PROCESS. The infinite WAIT must be present since otherwise there will be a path through the PROCESS without a sensitivity list or a WAIT, which is incorrect.

The 'system' for testing the AND gate is the ENTITY test_and with the ARCHITECTURE t_and. Two COMPONENTS are used, namely 2_input_and and clock_gen. Two instances of clock_gen, c_gen_1 and c_gen_2 are created. One of these will be set to run at half the speed of the other. Hence, if stop is set sufficiently large, all four possible inputs will be generated. Values of the GENERIC times must be supplied in the ARCHITECTURE t_and, since it is possible to configure the system without a CONFIGURATION section. The most general and flexible system of setting GENERICs is shown. The values set by the CONFIGUR-ATION con_t_and of the system will override those set in the ARCHITEC-TURE t_and. These values are shown as mark = space = 50 ns for c_gen_1 and 100 ns for c_gen_2. In both cases stop is set to 200 ns, sufficient for the four phases. The definition of 2_input_and in Prog. 5.3 is a general one, so the CONFIGURATION in Prog. 5.4 must set the actual values of t_{plh} and t_{phl}. Prog. 5.4 is one of several possible ways of achieving the required ends and is believed to be the most general.

Prog. 5.4. Test of a sub-circuit; a 2-input AND gate.

```
ENTITY clock_gen IS
    GENERIC (mark, space, stop : TIME);
    PORT (clock : OUT BIT);
END clock_gen;

ARCHITECTURE clk OF clock_gen IS
BEGIN
    PROCESS
    BEGIN
        IF (NOW < stop) THEN
            clock < = '0,' '1' AFTER space;
            WAIT FOR (mark + space);
        ELSE WAIT;
        END IF;
    END PROCESS;
END clk;

ENTITY test_and IS
END test_and;

ARCHITECTURE t_and OF test_and IS
COMPONENT clock_gen
        GENERIC (mark, space, stop : TIME := 1 NS);
        PORT (clock : OUT BIT);
END COMPONENT;
COMPONENT 2_input_and
        GENERIC (tplh, tphl : TIME := 1 NS);
        PORT (x, y : IN BIT; z : OUT BIT);
END COMPONENT;

SIGNAL a, b, c : BIT;

BEGIN
            c_gen_1 : clock_gen
            PORT MAP (a);
            c_gen_2 : clock_gen
            PORT MAP (b);
            gate : 2_input_and
            PORT MAP (a, b, c);
END t_and;

CONFIGURATION con_t_and OF test_and IS
        FOR t_and
            FOR c_gen_1 : clock_gen USE ENTITY WORK.clock_gen(clk)
            GENERIC MAP (mark = > 50 NS, space = > 50 NS, stop = > 200 NS);
            END FOR;
            FOR c_gen_2 : clock_gen USE ENTITY WORK.clock_gen(clk)
            GENERIC MAP (mark = > 100 NS, space = > 100 NS, stop = > 200 NS);
            END FOR;
            FOR gate:
                2_input_and USE ENTITY WORK.2_input_and(and_2)
                GENERIC MAP (tplh = > 15 NS, tphl = > 23 NS);
            END FOR;
        END FOR;
    END con_t_and;
```

5.5   More general modelling facilities

The VHDL PROCESS also permits the use of a number of other control structures. Some of these are probably of more use in designing element models than in modelling the setting of primary inputs to a network. Nevertheless, in the work of design validation, the ability to alter the course of a simulation under the control of the results being achieved is very useful. It is totally *useless* for testing purposes, of course.

5.5.1   WAIT

When simulating some structures, it is more convenient to set a sequence of actions going and then to wait for the network to reach some predefined logical state. For example, a test sequence for a counter might be written without knowing how long it would take to count through all its states. It is more convenient to let the clock run until the counter reaches its final state – say all ones. VHDL has a construct of the form

WAIT UNTIL < boolean expression >

where, in this case, < boolean expression > would be all the counter bits being '1'.

Such a test is extremely dangerous in this form. Remember that the intention is to check a design which may contain faults, whether in concept or in execution (typing mistakes!). Such a fault may well result in < boolean expression > never becoming true. To prevent the simulation running for ever a construct

WAIT FOR < time expression >

should be included. < time expression > is a time that is long enough to allow correct operation plus a bit to cover a small error in estimation of the time required and/or get a better view of a fault if one occurs. It should be small enough to prevent undue waste of computational resources when something is seriously wrong.

It is also possible to write

WAIT ON < signal list >

Any changes of any signal in < signal list > will cause the WAIT to expire. It is worth noting that a PROCESS with a sensitivity list has an implicit WAIT ON immediately prior to the END PROCESS; statement. The < signal list > is the PROCESS sensitivity list.

For full generality, a single WAIT statement can contain all three types of condition. It will expire as soon as any *one* becomes true.

5.5.2 The LOOP statement

An alternative method of generating a clock signal is the LOOP construct. Prog. 5.5 shows one example in which the clock is to run up to the user supplied time, stop. When testing networks such as counters it may be preferable to specify the number of clock cycles. Prog. 5.6 shows a possible fragment. The loop counter, i, is declared by its use in the LOOP construct and must not have any other declaration (in contrast to most programming languages). Note that i can be counted up or down as is convenient.

In some cases it is useful to be able to abandon a loop if some condition becomes true. VHDL provides two mechanisms for different purposes.

- NEXT WHEN <condition>;
  If the condition becomes true, the current iteration of the LOOP is abandoned and the next begun. It is effectively a jump to a place immediately prior to the END LOOP; statement.
- EXIT WHEN <condition>;
  This is effectively a jump to immediately after the END LOOP; statement, thereby abandoning the LOOP altogether. Prog. 5.7

Prog. 5.5. Use of the LOOP construct.

```
WHILE (NOW < stop) LOOP
        CLOCK < = '0,' '1' AFTER space;
        WAIT FOR (mark + space);
END LOOP;
```

Prog. 5.6. Alternative LOOP control.

```
FOR i IN 0 TO 15 LOOP      -- or i IN 15 DOWNTO 0 LOOP
        clock < = '0,' '1' AFTER space;
        WAIT FOR (mark + space);
END LOOP;
```

Prog. 5.7. Use of EXIT from a LOOP.

```
inner_loop : WHILE (NOW < stop) LOOP
                    clock < = '0';
                    WAIT FOR space;
                    IF (NOW > stop) EXIT;      END IF;
                    clock < = '1';
                    WAIT FOR mark;
                END LOOP;
```

causes the clock to finish exactly on stop regardless of the value of stop. Previous examples may behave unexpectedly if stop is selected injudiciously.

If LOOPs are nested then the EXIT statement shown above applies to the local LOOP only. However, a LOOP may have a label attached shown as inner_loop in Prog. 5.7. If each LOOP in a nested set has a label, then, by writing

EXIT <label> WHEN....
the exit can be made to apply to the (non-local) LOOP with the name <label>.

5.5.3 CASE statement

The CASE statement operates in a manner similar to that of any programming language. Prog. 5.8 shows a multiplexer with a control signal, c, data inputs a and b and output z. z is a copy of a if c is '0' and of b if c is '1.' Each signal is of type BIT. Notice that as c can take one of only two values, all the possibilities are enumerated. In the CASE statement all possible values of the selector (c here) must be covered and there must not be any duplicates.

The OTHERS construct is a catch all for a number of 'don't care' values of the selector, or where the result is the same. The comment in Prog. 5.8 is an alternative to enumeration of the '1' case of the selector. If a particular selector leads to no action at all, the NULL statement can be used. If the multiplexer used two selectors to select amongst four data inputs, care must be taken in the way it is written. If a temporary value is used as a decode of the two selectors then it *must* be a VARIABLE and not a SIGNAL. The reason is as given earlier, that a SIGNAL is predicted, and hence the PROCESS would use an old value of the decode to select amongst the data inputs, rather than the new value.

Prog. 5.8. Use of the CASE statement.

```
mux : PROCESS (a, b, c)
    BEGIN
        CASE c IS
            WHEN '0' = > z < = a;
            WHEN '1' = > z < = b; -- or OTHERS z < = b;
        END CASE;
    END PROCESS;
```

This chapter has given a flavour of the sort of facilities that are available in a language which seems likely to be the *de facto* world standard for some time to come. The emphasis of the chapter is on controlling simulations. The facilities can also be used in writing element models. The description in this chapter is far from being complete but should be a useful introduction to the implementation of the type of facilities that are necessary for controlling a simulation and for writing models.

6

Simulation algorithms

6.1   Introduction

There are two approaches to simulation.

- Simulate for functional correctness, ignoring all timing considerations, and then use a timing verifier to check that time constraints are met.
- Simulate in an environment in which the models include timing. As one can never guarantee that all paths through the logic have been exercised, it may still be advisable to use a timing verifier.

The simplest approach to simulation is to have a separate procedure for every logical element in the network being simulated, and the connections between the elements are then mirrored in the structure of the machine code of the program. The entire structure of the network is thus mirrored in the store of the machine doing the simulation. This takes up a great deal of storage space, but is very fast in running, since there are no lengthy lists to be searched and manipulated.

The amount of storage can be reduced by having only one procedure for each element type, and a small amount of storage for every element holding the element-specific data. In the previous description there is a copy of the procedure for every element which uses it and hence no procedure entry and exit as such. With only one copy each procedure requires a call. Procedure calls need machine states to be saved temporarily and restored on exit. This is expensive in CPU resources. Storage space is saved, since only one copy of each different procedure is needed. The cost is the time taken in procedure calling.

In the simplest simulator, all elements will have the same delay. The number of gate evaluations can be considerably reduced by first examining the structure of the network and putting gates in an appropriate order

(Section 6.2). In this order it is guaranteed that, for a combinational network, no gate will be simulated more than once per step. Special arrangements are required to handle feedback. The particular case of flip-flops is treated by regarding them as basic elements rather than as devices with feedback.

This type of simulation is only useful for checking functionality, since all element delays are the same. Since real gates have delays which depend on the direction of change of the output ($0 \rightarrow 1$ or $1 \rightarrow 0$), to include real timing would be impossible. It would mean that ranking would be dependent on the data. Instead one constructs a set of 'event lists,' one for each time interval. The details are given in Section 6.3. Briefly, when a change of gate output is computed at time T, say, the change is placed in the list for time $T + \delta t$, where δt is the gate delay. Lists are emptied in time order. When items are taken off the list a second table is consulted to find which elements are driven by this output.

The first type of simulator is designed to check logical correctness at maximum speed. Timing accuracy is regarded as of little importance – it will be checked in detail at a later stage in the design. It is an **oblivious** simulator, as it evaluates every element whether it is essential or not. It is usually run by compiling the network and is frequently known as the **compiled code** simulator, since the structure is reflected in the machine store and program. It is a static structure, allowing no timing detail to be seen. Oblivious and compiled code do not necessarily go together. Following convention, this type will be referred to as compiled code.

The most complex of the above simulators enables detailed timing to be examined. It is based on sets of tables. The structure of the system being simulated is reflected in some of these tables, and the dynamic timing in others. It is referred to as a **table based event driven** simulator (or table based or event driven for brevity). Such a simulator usually (but not necessarily) is run interpretively.

Although a continuum of simulator designs is possible, and simulators representing several points in the continuum have been implemented, this book will limit itself to the simplest and the most complex as the basis of the most important simulators in use. It will be seen that other features can also be included in the event driven simulator, and that both compiled code and event driven simulators can be useful when properly used by the designer. It is important that the designer understands the strengths and weaknesses of the tools available. As indicated earlier, they are *aids*, not gurus.

There are other distinctions between the two. The compiled code approach presumes a synchronous system is being simulated – that is a system consisting essentially of blocks of combinational logic separated by clocked registers. The event driven approach can handle any system, including asynchronous systems. In practice almost all systems have some asynchronous connection, so special techniques are needed to handle these in the compiled code case.

Section 6.2 describes the compiled code simulator in more detail and Section 6.3 describes the event driven simulator. In each case a detailed example is included. The reader may omit the examples if that detail is not of interest.

6.2   Compiled code simulation

6.2.1   Basic procedures

The essence of a compiled code simulator is that the structure of the system being simulated is reflected in the computer store, and that each logical element has its own code. Timing is ignored in the simplest form. Logical correctness alone is checked. Once the designer is satisfied, some timing checks are made, usually with a different piece of software (timing checks may be made earlier to find gross errors).

Consider the simple not equivalence circuit of Fig. 6.1. Suppose that, due to the way the designer entered the data into the computer, G2 was to be simulated first. Suppose the test architecture causes A to change from *1* to *0*. The output of G2, D, is evaluated. Some time later G1 is evaluated. If B is *1*, then C changes, and it will be necessary to recalculate D.

Since all delays are the same and detailed timing cannot be evaluated, evaluating elements several times is wasteful. A reduction in work can be achieved by a simple **rank ordering** of the gates.

To rank order the gates, proceed as follows, Fig. 6.1 being used as an example.

Fig. 6.1. Simple example for simulation.

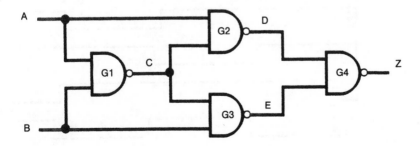

- Primary inputs are assigned to level 0.
- Find all gates with inputs connected *only* to primary inputs. G1 is the only such gate in the example, and is assigned to level 1.
- From the remaining gates, find those whose inputs are connected to levels *0* and/or *1* only. G2 and G3 are assigned to level 2.
- Proceed further in this manner. At each step find gates not yet assigned whose inputs are all assigned to a level. The level to which a gate is assigned is one greater than the highest level of any of its inputs.

Fig. 6.2 shows waveforms for some input changes as a result of simulating Fig. 6.1 with the rank order just derived. Delays of one unit per rank are shown in order to indicate the ranks. The model of the gate is a two-input NAND. The inputs A and B are initially 0. The level 1 gate(s) is simulated, giving a *1* at C.

G2 is simulated next, followed by G3 (G3 followed by G2 is equally good). Each gives a *1* out (D and E). Finally G4 is simulated, giving a *0* at Z.

A now changes to *1*. G1 is simulated, but, as B is still at *0*, C does not change. Next consider G2. Both inputs are now *1*, so D becomes *0*. Simulation of G3 shows no change at E. Finally, in level 3, G4 is simulated. D is now *0*, so Z becomes *1*.

The reader should repeat this simulation for the remaining part of the waveforms of Fig. 6.2.

It will be seen later that, with real timing, it is possible for a narrow pulse to occur at D or E. This does not invalidate the use of the compiled code simulator. Indeed, as a check on the logical correctness of the design, and

Fig. 6.2. Waveforms for Fig. 6.1.

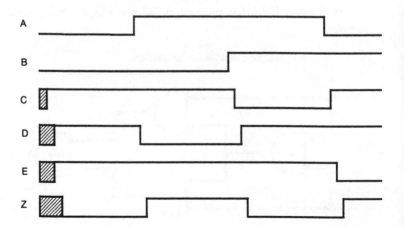

when used with a timing verifier (Chapter 8), it is particularly useful. Modifications to the technique may find some short pulses but not all of them. That is worse than none at all, since the user is often misled into thinking that the checks are comprehensive. When there are no checks then there is less likelihood of that. Thus the simplest technique is the best. This simulator is known as a **levelised compiled code** simulator, or **LCC**.

The next two sections give some details.

6.2.2 Simulator structures

As with any simulator, data describing the interconnection of the network must be held in a suitable memory. For each logical element it will be necessary to hold data on inputs and outputs. It is preferable to hold signal data in a separate table, so the element inputs and outputs will be in terms of addresses in the table describing the network. Hence, if a signal is an output of one element and the input of two others, it is recorded only once and has to be modified in only one place when a change occurs, not in three. For generalised logic elements this could be a very large number of addresses – several hundred. As other logical elements might require only two (e.g. an inverter), making every logic element fit the same sized piece of storage would be very wasteful. Further, whatever number was chosen it is certain that some user would want a few more.

There are two solutions to this. The first is to have a large pool of storage and let each element have a list of addresses. The lists must be ordered in a form known to the routine which is to compute the element output, the **evaluation routine**.

A second solution is to restrict the size of the logic elements. In the Yorktown Simulation Engine (YSE) (Pfister 1982, Denneau 1982, Kronstadt and Pfister 1982; Denneau *et al.* 1983), a simulator built in hardware by IBM, the elements are limited to four-input one-output devices. Other simulators allow only three inputs. Libraries of more complex elements in terms of these basic devices are available. Thus only five addresses are needed per element in the YSE and all are provided for every element. There will still be some wastage, but it will be strictly limited. The YSE structures can equally well be put in software and are a useful example of how a compiled code simulator can be constructed.

Each element in the network can be represented as a record holding

- the address of inputs and outputs
- the function.

This is described in the YSE as the **instruction memory**.

The connectivity of the network is implied in the addresses of the signals.

The elements must be placed in an order implied by the levelising process. Hence the first element evaluated must be at level 1. The levelising process implies that the output of this element cannot be used in level 1. More generally, the output of an element at level N cannot be used in any level less than $N+1$. Thus the order of evaluation within a level does not matter. The input values used in evaluating an element in level N will always be the up to date values from level $N-1$ or earlier.

A second block of memory is assigned to values of the signals. This can be in any order. The network compiler will ensure that connectivity of the network is implied by signal addresses in the 'instruction memory.' These signal addresses refer to the **data memory**.

6.2.3 Detailed example

Table 6.1 shows a possible structure of the instruction memory for Fig. 6.1 and Table 6.2 shows the data memory. Table 6.1 is arranged in level order as indicated in the comment column. Primary inputs A and B appear first as they are at level 0. The first line for A contains an address in the data memory, 5, where the value of signal A is to be found and a flag bit set to 0. The flag implies that the next line is the function of the element, in this case primary input. No evaluation is required. The function line is always the last line for the element, implying that the next line is a new element if there is one. The next two lines of the instruction memory are similar for B. Following that are four lines relating to gate G1. The first line is the address, 3, where the value of the output is stored in the data memory. The flag is a *1*, implying that the next line also belongs to G1. Single output elements are assumed in this description, so the second line contains the address of an input, as does the third line. As an aid to the reader, the comment columns in Table 6.1 indicate the signal in Fig. 6.1 to which the line in the table refers. The flag in the third line of G1 is *0*, indicating that the next line is the last for this gate. The function in that line is two-input NAND. The remaining gates follow similarly.

In the data memory (Table 6.2) it is assumed that all signals are initialised to the 'unknown' value X. The evaluation process starts at the beginning of the instruction memory. The primary inputs are obtained from a VHDL PROCESS in a test bench architecture (Fig. 5.1), or from an equivalent external source. Address 1 of the instruction memory is read and address 5 of the data memory set to 0. Similarly with B, data memory address 1 is set to 0. Table 6.2 shows this at period 1 level 0. The function is 'primary input' in both cases, so no evaluation is needed.

The next read of the instruction memory is for G1. The input addresses in the data memory are seen to be 5 and 1, which are both read. They are 0.

Table 6.1. *Instruction memory for Fig. 6.1. Note: blank lines for readability only*

Level	Comment Element/signal	Data Addr	Flag
Level 0	Input A	5	0
	function	PI	–
	Input B	1	0
	function	PI	–
Level 1	G1 output C	3	1
	input A	5	1
	input B	1	0
	function	NAND_2	
Level 2	G2 output D	2	1
	input A	5	1
	G1 output C	3	0
	function	NAND_2	
	G3 output E	4	1
	G1 output C	3	1
	input B	1	0
	function	NAND_2	
Level 3	G4 output Z	6	1
	G2 output D	2	1
	G3 output E	4	0
	function	NAND_2	

Table 6.2. *Data memory for Table 6.1 and Fig. 6.1. Only changes shown for clarity*

Address	Signal	Period 1 level				Period 2 level				Period 3 level				Period 4 level			
		0	1	2	3	0	1	2	3	0	1	2	3	0	1	2	3
1	B	0				0					1			1			
2	D	X	1					0				1					1
3	C	X	1					1				0				1	
4	E	X	1					1				1				0	
5	A	0					1				1				0		
6	Z	X			0				1				0				1

The gate is evaluated according to the function (NAND). This is usually a simple look-up table for an LCC simulator. The instruction memory indicates that the output of this gate (C) is at line 3 in the data memory, and the value is stored there (Table 6.2, period 1 level 1).

The system then steps through the instruction memory. G2 inputs are in lines 5 and 3 of the data memory and have the values *0* and *1* respectively. The output, D, is *1*, therefore, and is placed in line 2 of the data memory (address read from G2 output line in Table 6.1). Subsequent gates are evaluated with the results shown in Table 6.2. When the system reaches the end of G4, all elements have been evaluated once and once only. A new set of primary inputs is now set and a new clock period commences. Table 6.2 shows the effect of three changes of the input, A to *1*, B to *1* and A back to *0* as shown in Fig. 6.2.

6.2.4   Handling feedback

One problem that has to be solved is what to do with feedback. Consider the simple R–S flip-flop shown in Fig. 6.3. The signals R and S are both level 0. G1 is at a level one greater than the level of G2 output. G2 is at a level one greater than the level of G1 output. This is a deadlock situation.

The way to deal with this is to break the feedback loop. Suppose that the second input of G1 is disconnected from Q and treated as a primary input at level 0 and with unknown logic value. The feedback loop is now broken as shown in Fig. 6.4. When the circuit is simulated, G1 is evaluated at level 1,

Fig. 6.3. Simple R–S flip-flop.

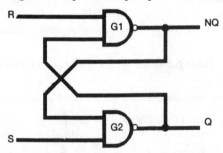

Fig. 6.4. R–S flip-flop redrawn with feedback broken.

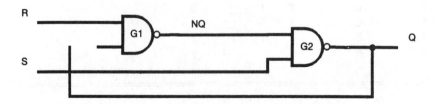

Table 6.3. *Simulation of Fig. 6.4*

Signal	Period 0	Period 1 level 0	1	2	Period 2 level 0	1	2	Period 3 level 0	1	2	Period 4 level 0	1	2
S	*1*	*0*	*0*	*0*	*0*	*0*	*0*	*0*	*0*	*0*	*1*	*1*	*1*
R	*1*	*1*	*1*	*1*	*1*	*1*	*1*	*0*	*0*	*0*	*1*	*1*	*1*
G1 i/p	X	X	X	X	1	1	1	1	1	1	1	1	1
Q	X	X	X	1	1	1	1	1	1	1	1	1	1
NQ	X	X	X	X	X	0	0	0	1	1	1	0	0

Table 6.4. *Alternative levelising – both feedbacks cut*

Signal	Period 0	Period 1 level 0	1	Period 2 level 0	1	Period 3 level 0	1	Period 4 level 0	1	Period 5 level 0	1
S	*1*	0	*1*	*1*	*1*	*1*	*1*	*1*	*1*	*1*	*1*
R	*1*	0	*1*	*1*	*1*	*1*	*1*	*1*	*1*	*1*	*1*
Q	X	X	1	1	0	0	1	1	0	0	1
NQ	X	X	1	1	0	0	1	1	0	0	1

and subsequently G2 at level 2. Table 6.3 shows such an evaluation. S and R start at *1* and Q and NQ are unknown, *X*. Primary inputs are shown bold.

Suppose S now goes to 0. Level 1 is not affected, but at level 2, Q is forced to *1*. As the feedback loop is broken, nothing further occurs. However, note that the inputs to G1 are now known, so further simulation should take place. Thus *the feedback signal is a level 0 signal.* Another cycle of simulation takes place causing NQ to become *0* in level 1. There are no changes at level 2. The simulation *does* take place.

In Table 6.3, period 3 shows what happens when R goes to 0. NQ is forced to *1* at level 1, and, as S is already *0*, Q remains at *1*, so there is no change at level 2. Again, the simulation takes place.

In period 4, R and S both return to *1* at the same time. At level 1 NQ is driven to *0*. This zero causes Q to remain at *1* in level 2. This will always be the case with this simulator. However, in a real circuit it is not known which state the flip-flop would assume. *The user is warned that such misleading results can be a feature of simulation.* Part of the test program generation is ensuring that such circumstances are properly separated and tested.

An alternative levelising is to break both feedback paths and regard both gates as level 1. Table 6.4 shows the effect when S and R are both *0* and go to

1 simultaneously. In period *1*, Q and NQ become *1*. S and R now go to *1*. As all gate inputs are *1*, Q and NQ go to *0*. Both gates now have a zero input, so outputs change to *1*. An oscillation is set up.

In *all* simulators there is a danger of oscillation in networks involving feedback. This example shows that the LCC simulator can handle matters reasonably if care is taken. It would seem that a feedback loop should be broken only once, but there may be many loops in a large system, and all must be broken. There may be problems ensuring that feedback loops are broken just once in such systems.

This hints at a problem with any simulator. If the logic is correct, the feedback loops can be broken. Oscillations should be detected under some input conditions if a loop remains unbroken. Such conditions may result from a fault in the design which introduces feedback where it was not intended – for example as the result of a misspelt signal name on data entry. This is clearly a method of finding design faults. However, these faulty oscillations must be distinguished from the proper oscillations of clock signals.

Arguments such as those above lead to the conclusion that LCC simulators should only be used with synchronous systems such as that illustrated in Fig. 6.5. Signals on the input to R_a are assumed to have settled. A clock signal (or its effect) moves these values to the output of the register as a level 0 simulation. The combinational logic, Y, is now simulated in levelised form. Y must contain no feedback loops. Once complete, a second 'clock' moves the output of the combinational logic to R_b, and new inputs to R_a. Feedback from R_b to R_a is allowed, of course, since the registers have the effect of breaking the feedback loop during any one phase of the simulation.

Fig. 6.5. A synchronous system.

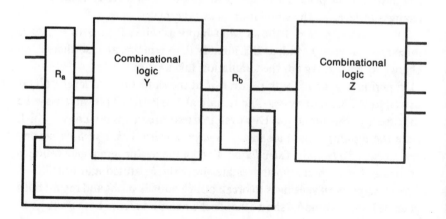

If the registers are latches (pass data when the 'clock' is in one state, stop it when the 'clock' is in the other state), some modification to this structure will be necessary, but the simulation will then work. If some blocks of logic are known to be stable for half of the clock period, it may be possible to reduce the amount of computation by separating memory into two parts and only scanning one part in each half clock period.

The problem is that no system is completely synchronous. This leads to a need for very careful examination of the asynchronous parts by a skilled designer, or a need for simulation by a different type of simulator. In spite of this, the LCC simulator is a very powerful tool, and is in full use by very large sections of the industry.

The compiled code simulator runs very fast, since the code traces the links between elements, and time is ignored (the significance of this remark will appear later). Speed can be further improved by removing buffers and inverters at 'compile' time, since buffers have no logical effect, and evaluations use boolean equations which can take into account any inversions. On the other hand, compilation times can be quite long and must be included in the run times. Maurer (1991) claims 70% of the time is in compilation. Any change in network or devices requires a full recompilation. This is not true of the event driven simulator (see below).

6.2.5 Some comments

At this stage it is as well to be aware of several matters.

Every gate is evaluated every time any level 0 signal changes value. For synchronous logic this corresponds to the activity in one clock period of the system. Where the setting of a register causes no change, it might be possible to reduce the work by not evaluating the following block of logic. However, this would add to the control of the simulation and might save little. After all, a system in which a block of logic is idle for a clock period is designed to run inefficiently. On the other hand, workers at Hitachi (Takamine *et al.* 1988) have suggested that 'clock suppression' can result in considerable savings with an event driven simulator (see next section). With multiphase latched systems, simulating a section of logic where the clock is inactive is clearly wasteful of resources.

A 'time period' as applied to an LCC simulator generally refers to the clock period of the synchronous logic and not to the simulation of one level. As every logical element is evaluated in this time period, the **activity** is said to be 100%. Indeed, in one design where there was some limited and strictly controlled feedback, the authors claimed the activity was more than 100%! (Wang *et al.* 1987). When the event driven simulator is considered in Section 6.3, real timing is introduced. Time steps are *much* less than a clock period, and hence the number of elements that require evaluating is much less than 100% – typically 1% or less. However, for the same piece of logic and same test patterns, there are many more time steps. Many authors are unclear on this point and frequently make improper comparisons. In comparing the two methods of simulation, one must be clear as to what is being compared.

6.3   Event driven simulation

6.3.1   Introduction

A compiled code simulator models the logical operation of a system, but ignores the timing. The models for elements can be very simple, since the problems of precise timing and events overtaking one another (see Chapter 7) do not arise. An event driven simulator is capable of modelling time at a very detailed level. In a compiled code simulator, a 'time step' is equivalent to a clock period. In a clock period the activity can be very high. The strength of an event driven simulator is its ability to provide detailed timing. The timing units will be fractions of a gate delay. Hence there are many time steps in a clock period, and the number of gates requiring evaluation in each of these steps will be much less than those requiring evaluation in a full clock period. Activity of less than 1% is quite normal, and 20% would either be very high or a rather special design.

A compiled code simulator evaluates every logical element at every time step regardless of whether there is an input change or not. Evaluating every element at every time step with activity of, say, 5% is very wasteful. An event driven simulator evaluates a particular element 'on demand' – only when an input changes. However, the evaluation of a gate with in-line code in an LCC simulator is very fast. The event driven simulator has to extract data from tables, so many more machine instructions are needed. Even with the same gate models as an LCC simulator, the activity needs to be less than about 1% for the event driven simulator to be faster. It must be emphasised that the primary advantage of the event driven simulator is *not* speed, but timing accuracy. In practice, the gate models are much more complex than those needed for zero or unit delay simulation of the LCC simulator. In these circumstances, the notion of unit delay event driven simulation as proposed by a number of authors is not very sensible, at least in the opinion of this writer. Someone quoted a speed gain for unit delay simulation in an event driven simulator of less than a factor of two.

Fig. 6.6 outlines the operation of an event driven simulator. The following describes what happens without explanation. The explanation will appear in subsequent sections.

The simulator assembles potential output changes or **events** in order of the time at which they are expected to occur. At a given time, T, the simulator first examines any external drive for the primary inputs and adds drives for the current time to the events in the main event list. Thus primary inputs are treated as outputs of pseudo-logic elements as in the test bench of Fig. 5.1.

The simulator then works with the set of events for time T. An event is extracted from the list. The event data includes an address in the fan-out table. Starting at this address, the fan-out table first contains a pointer to the output driver. The value of this output signal is updated. Subsequent fan-out table addresses point to element inputs driven by the current event. For each of the driven elements in turn, the data is extracted from other tables, the element type is determined and the model evaluated. Outputs will be predicted to occur some time in the future, δt say, and these will be added to the event list for time $T + \delta t$. It will be seen later that these will be passed on even though there may appear to be no change. For multiple output elements, there will be multiple predictions, each with its own value of δt. For different element types, the values of δt will also vary. Thus there will be many different values for δt, so there must be many sets of events; one for each value which δt may assume.

The procedure is repeated for each event at this time. When all have been

Fig. 6.6. Outline flow of an event driven simulator.

```
Time = T = 0
Repeat
        While (events for time T in test program)
              place event into main set for time T

Repeat (with set of events for time T)

        extract event
        Repeat
              find a fan-out from fan-out table
              extract data for affected element
              evaluate affected element model
              Repeat place predicted output in set for time T + δt
              Until no more outputs
        Until no more fan-outs

Until no more events

Increment T

Until (all event sets are empty AND end of test program)
        OR run out of computing time.
```

processed, the current time frame is ended. Time is incremented and the process repeated until the end of the simulation. The reader may find it useful to refer to Fig. 6.6 from time to time in what follows.

6.3.2   Basic procedures

The outline structure of an event driven simulator is shown in Fig. 6.7. There follows a description of the operation in general terms. An example with some typical figures for a small circuit will be found in Section 6.4. Readers might find it helpful to re-read this section in parallel with Section 6.4.

The **network memory** contains the state of the network being simulated in order of network element. Each record in the structure contains the state of all inputs and outputs of one logical element together with a reference to the relevant model. When the input of a particular element changes, all the data relating to that element is passed to the evaluation routine for that element. This data includes the 'new' input and output states, the 'old' states which are not being changed, and possibly those which are. It also contains a unique identifier for each output known as the **fan-out index, FOI**. This will be used later as an address to a table, the **fan-out memory, FOM**, containing information on where to find the inputs which are driven by that output.

The effect of an input change is evaluated by the evaluation routine and output predictions forwarded to the **event memory**. The model being

Fig. 6.7. Basic event driven simulator.

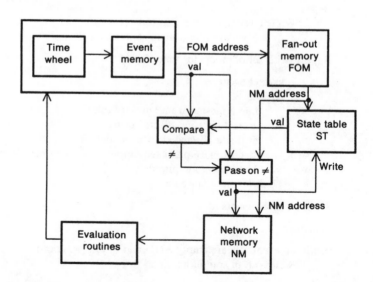

evaluated knows the delay from input to output for the element. This may be different for different input and output combinations. The delay, δt in the previous paragraph, is forwarded to the event memory.

All output predictions must be forwarded to the event memory, even if, at this stage, there appears to be no change. Consider the logic element and waveforms shown in Fig. 6.8. At time 0 input A goes to *0*, and the evaluator predicts that the output will change to *0* at 12 time units ahead, say. This prediction is placed at time $T+12$ in the event memory lists. At time 5, input B goes to *1*, and a prediction that the output will change to *1* after 9 time units is made, i.e. at time 14. The output at time 5 is already *1*, and it is tempting to throw away this prediction. However, a prediction for a change to *0* has already been made, but is not known to the evaluation routine, since this is shared with other elements of the same type. If the prediction made as a result of B changing is thrown away, the change of the output to *1* at time 14 will be lost and the final state of the output will be *0*, incorrectly. Thus all predictions must be sent to the event memory regardless of the current state of the output. The reader will find many descriptions of event driven simulation in which events are forwarded to the event memory only when current and predicted outputs are different. *Such a simulator will not handle the situation just illustrated.*

Note that there are alternative ways of achieving this effect. One might be to place time markers in the network memory. However, the mechanism above can be used for other purposes and so the author believes it is the more general and useful method.

Fig. 6.8. Need for prediction at time 5.

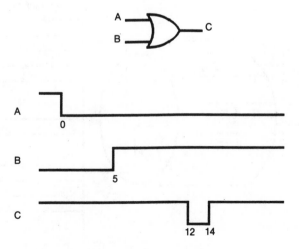

The event memory is a list of all predictions in time order. This could be a single linked list in which records contained fields for value, fan-out index and absolute time. To insert a prediction, the time fields of the list would be searched to find an appropriate place to insert the new event. However, the total number of events in the memory for a reasonable sized piece of logic will be very large, and hence finding the place at which to insert an event could take a long time.

The usual approach is to form a list of lists in time order, Fig. 6.9. The first list is a set of pointers to the event memory. There is notionally one pointer for each time interval from current time plus one. This could be a very large number. For any given simulation, once can assume that the vast majority of element delays are less than some number of time intervals – say 1000. Thus one might limit the number of pointers to 1024, and re-use them at every 1024 time steps. The list of pointers is thus a circular list, and is called a **time WHEEL** or **chronwheel**. The wheel itself is an array of pointers and hence is accessed as indicated in Fig. 6.9. Searching is not required. Clearly provision needs to be made for the occasional very long delay. This will be discussed later.

Each pointer in the time wheel indicates the address of a list of events that has been scheduled for that time, Fig. 6.9. Each list must be long enough to hold the maximum number of events for any time interval. This cannot be predicted. Furthermore, the lists for the times furthest ahead are likely to have relatively few events. This is a very poor use of store and would reduce

Fig. 6.9. Time wheel and event memory.

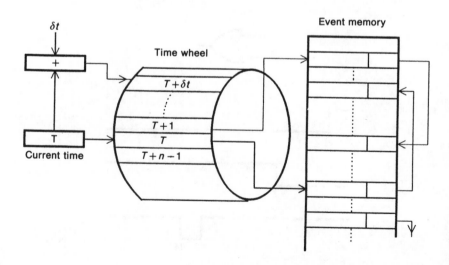

the size of the network that could be simulated with given resources, or increase disc thrashing and hence simulation time.

An alternative is to use a linked list for each time slot in the time wheel. Each record in this list can be drawn from a large pool of memory. It is only necessary to provide sufficient store for the sum of all the event lists, and there is much less wastage.

The event memory consists of a series of records as shown in Fig. 6.10. Each record holds the predicted value of a signal, the fan-out index (FOI) and a pointer. When an element evaluation wishes to schedule an event, it sends data to the event memory. The delay, δt, is added to current time, T, and the pointer at $T + \delta t$ read. The new event is added to the list indicated by the pointer. If this pointer is *null* (there are no events scheduled for this time slot as yet) it picks up a 'free' event memory address from a list of unused locations (**'free list'**). Suppose the location found from the free list is 31 (Fig. 6.11(a)). The predicted value of the signal, val(a), and its fan-out

Fig. 6.10. Format of an 'event' in the event memory.

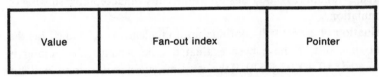

Fig. 6.11. Updating the time wheel and event memory.

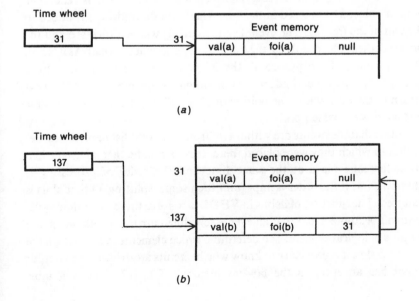

index, foi(a), are placed in record 31. The pointer of the event memory record is set to *null* and the time wheel pointer set to the event memory address, 31.

Fig. 6.11(*b*) illustrates the case where the pointer in the time wheel is not *null*. It is 31, and points to the line in the event memory containing val(a) and foi(a). The pointer of record 31 is *null* as this is the only event in this list. A new event for this time is to be scheduled. The next location in the free list is 137, say. The pointer, 31, is placed in the pointer field of line 137. Line 137 now points to line 31. val(b) and foi(b) are placed in the other fields of line 137. The time wheel pointer is set to 137. There are now two events in the set for this time, the new event having been added to the *front* of the list. The list is followed from the time wheel value of 137. The record 137 holds the pointer 31, and the record 31 has a *null* pointer indicating the end of the list. The order of events in a particular time slot is not important.

The reason for keeping a free list is that the events are placed in successive locations in the event memory as they arrive, but these are not ordered in time. When events are removed in time order, the free locations are scattered at random. To tidy this up would cause unacceptable slowing of the simulation.

Evaluation of events will continue until all elements scheduled for the current time slot, T, have been evaluated. The current time pointer is incremented to $T+1$ (Fig. 6.9). If the new time wheel pointer is not *null*, the event memory location to which it points is read. Return to the example of Fig. 6.11 and recover the event from the time slot previously filled, T' say. The time wheel pointer is 137. The record at location 137 is read. The pointer value, 31, is placed in the time wheel and the address 137 added to the end of the free list. The event [val(b), foi(b)] will be processed and the program then returns to see if the time wheel pointer at T' is *null*. As it is not, the (a) event will be processed, the T' pointer set to the pointer from location 31 – *null* – and address 31 returned to the free list. On the next return to the time wheel, the pointer at T' is *null*, so all events for this time slot have been passed on.

Notice that the events are withdrawn from the event list in reverse order to that in which they were placed there. It is axiomatic that all events in a given list take place at the same time and the order of processing is unimportant. If the order is important then some 'splitting' of time slots is required. The function of deltas in VHDL is to force these extra time slots.

It will be appreciated that all events scheduled are for signals which are outputs of logical elements. To determine which elements to evaluate in the new time slot, it is necessary to know which circuits are driven. Each output signal has an entry in the fan-out memory, Fig. 6.7. This is a table containing details of where the output is in the network memory. Fig. 6.12 illustrates this. A NAND gate drives two loads, a flip-flop and a logic block called FRED. Suppose the record in the network representing the NAND gate begins at address 258. The output is, say, the third entry in this record, so is at location 258+2 (not 3!), designated 258.2 in the diagram. The fan-out index of this output is 567, say. This is an address in the fan-out memory. On reading line 567 of the fan-out memory the address 258.2 is found. It will be appreciated from the example of Fig. 6.8 that the output of the gate could not be changed at the time of evaluation. This entry in the fan-out memory enables that change to be made.

A flag in the fan-out memory is set, indicating more data relevant to this signal. The next line has value 9356.7, which is the eighth entry of record 9356, probably 9363, of the network memory. This is the input of the circuit element FRED. The third entry, line 569 of the fan-out memory, indicates that this NAND gate also drives the fifth entry of record 83, probably 87, of the network, the D-input of a flip-flop. The flag of this line is 0, indicating no more fan-outs of this NAND element.

Notice that this table holds the connectivity of the network.

It will be realised that the number of events leaving the fan-out memory for the network memory is larger than the number arriving. Conventional wisdom suggests that each output drives 2.5 inputs on average, so for each event in there are 3.5 sets of data out, for the output and 2.5 inputs.

Fig. 6.12. Fan-out memory.

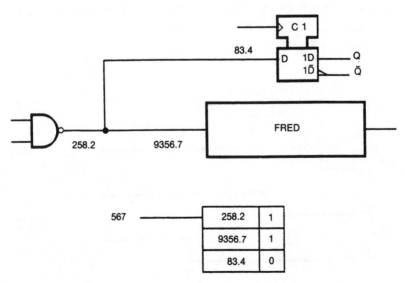

The **state table** (Fig. 6.7) holds the current state of all outputs, and is a copy of those in the network memory. It is shown in Fig. 6.7 as a separate table for convenience, but in most software implementations the data structures would be constructed in such a way that there would be only one copy, that in the network memory. When an event reaches this point, the outputs are actually changing. Referring back to the example of Fig. 6.8, at time 12 the state table value of C will be *1* and the predicted value *0*. At time 14 the values are *0* and *1* respectively. Hence both events are real ones.

The situation where the result of the event at B in Fig. 6.8 causes a fast response and cancels the event due to the input A is a more complex matter and will be discussed fully in Section 7.3. The alternate case of the use of the state table is illustrated in Fig. 6.13.

Input A changing to *1* causes a prediction that C will change to *1* at time 12. At time 5, B changes to *1*, and causes a prediction that C will change to *1* at time 17. Note that, if the evaluator had made a comparison with the output, this would still be seen as a change since the output change due to A has not yet been recorded in the network memory because it could be cancelled. At time 12, the first change is propagated to the elements driven by C, which are evaluated since C in the state table, *0*, is different from the predicted event, *1*. At time 17, the same predicted change is again found from the list in the event memory for this time, but all driven elements have already been evaluated for this value of input, and do not need to be evaluated again as a result of this event. The output signal value coming from the fan-out memory is compared with that in the state table. If they are the same, there is no need to pass this event on. This not only saves the average 2.5 evaluations. As all predictions from the evaluator must be

Fig. 6.13. Use of state table.

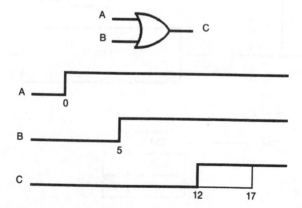

passed on, it also saves this happening, and avoids an explosion of unnecessary work. This is a factor of 2.5, since once activity has built up, the number of predicted events must be the same as those passed on or the situation of all inputs active or no inputs active will soon be realised.

If the output prediction is different from the value in the state table, the event is passed on to the evaluation routines. The signal value in the state table is updated. This is, therefore, a good place to detect changes and report them to the output process and hence the user, since it is at this point that predictions of *potential* changes are confirmed as *real* changes.

6.4   An example – a four-gate not equivalence circuit

Consider the simple not equivalence circuit shown in Fig. 6.14. There are just four two-input NAND gates and two primary inputs, A and B. The gates are numbered in an arbitrary way. There are four memories of interest (Fig. 6.7). These are each allocated an area of the real machine memory. Addressing of the memories will be written in the form of base and offset. For ease of reference, the first address, or base, in each memory will be given in literal form, namely

NM	network memory	e.g. 10 000
EM	event memory	e.g. 20 000
FOM	fan-out memory	e.g. 30 000
ST	state table	e.g. 40 000

To find a particular item of data the offset is added to the base to give the real address. Thus NM243 will be address 10243.

A 'network compiler' will construct the various tables. G1 is assigned to line NM1 of the network memory as shown in Table 6.5, and consists of five lines (compare with Table 6.1). The first contains the base address of data relating to G1 in the fan-out memory as described below (the fan-out

Fig. 6.14. Simple not-equivalence circuit (Fig. 6.1).

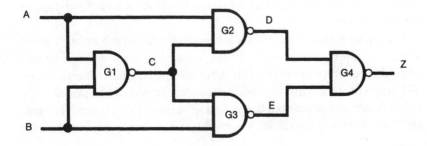

Table 6.5. *Network memory for example of Fig. 6.14*

	Network memory				Comment
Address	Data	Flag	Gate	Signal	Other
NM 1	FOM5	*1*	G1		data = FOI
NM 2	*1*	*1*		C	
NM 3	*0*	*1*		A	
NM 4	*0*	*1*		B	
NM 5	NAND_2	*0*			data = gate type
NM 6	FOM8	*1*	G2		data = FOI
NM 7	*1*	*1*		D	
NM 8	*0*	*1*		A	
NM 9	*1*	*1*		C	
NM10	NAND_2	*0*			data = gate type
NM11	FOM10	*1*	G3		data = FOI
NM12	*1*	*1*		E	
NM13	*1*	*1*		C	
NM14	*0*	*1*		B	
NM15	NAND_2	*0*			data = gate type
NM16	FOM12	*1*	G4		data = FOI
NM17	*0*	*1*		Z	
NM18	*1*	*1*		D	
NM19	*1*	*1*		E	
NM20	NAND_2	*0*			data = gate type

index). The second contains the output value and the next two contain the
input values. Putting the output value before the input values has
advantages for multiple output elements, but is not of the essence here. The
last line contains the element type (or a pointer to the evaluation routine for
this element). G2 data follows, occupying lines NM6 to NM10 and the
others follow on. The outputs are also assigned to lines ST1 to ST4 of the
state table (see Table 6.8 later). The flag bit is always set to *1* except in the
last line of data for an element. Thus, in lines NM1 to NM4 of G1, the flag
indicates that there is more data relating to G1. The flag of *0* in line NM5
indicates that this is the last line of G1 and line NM6 will be the first line of
the next element.

The inputs A and B are set by a process in the test bench architecture in
VHDL or can be considered to be the 'outputs' of some external circuit.
They also need an entry in the state table and are assigned to lines ST5 and
ST6 respectively in quite an arbitrary manner. They could be placed in the
network memory as well, but that would serve no useful purpose. Of course,
the state table is not essential, since gate output data is duplicated in the

Table 6.6. *Fan-out memory for the example*

Address	Base	Fan-out memory		State table address	Comment
		Data offset	Flag		Signal
FOM 1	NM 1	2	*1*	ST5	A
FOM 2	NM 6	2	*0*		
FOM 3	NM 1	3	*1*	ST6	B
FOM 4	NM11	3	*0*		
FOM 5	NM 1	1	*1*	ST1	C
FOM 6	NM 6	3	*1*		
FOM 7	NM11	2	*0*		
FOM 8	NM 6	1	*1*	ST2	D
FOM 9	NM16	2	*0*		
FOM10	NM11	1	*1*	ST3	E
FOM11	NM16	*3*	*0*		
FOM12	NM16	1	*0*	ST4	Z

network memory. If there is no state table then the primary input data must be in the network memory.

Arbitrary values such as 0 might be assigned to all values in the circuit. Those shown in Table 6.5 assume that A and B are set to 0 and a 'simulation' has been run to initialise the gate outputs. Thus C, D and E all become 1 since one input is 0, and Z becomes 0 since both inputs are 1. The input values are also appropriately set.

Table 6.6 shows the fan-out memory (FOM). Primary input A has arbitrarily been assigned to line FOM1. The two pieces of data in line FOM1 are a base and offset to the place where A is used in the network memory. Thus the first fan-out of A is to be found in the network memory at line $NM(1+2) = NM3$ (see Table 6.5 comments). The flag is 1, indicating another fan-out of A, which is found in the network memory line $NM(6+2) = NM8$. The flag now indicates that this is the end of the data for signal A. The data in the state table column gives the address of the signal in the state table, which was specified above to be ST5 for signal A. Other lines of the fan-out memory can be interpreted similarly. Thus, for G1, the output $\bar{C}$ is on line $NM(1+1)$, and the two fan-outs on lines $NM(6+3)$ and $NM(11+2)$ of the network memory.

It will be seen that the network memory describes the logical elements and the fan-out memory their interconnection. The network can be reconstructed from these two tables. The element type indicates to the evaluator which evaluation routine to use. The evaluation routine knows the order of the data in the network memory and what to do with it.

The data structures used would be easier to handle if they were single lines for each element. However, with bigger elements the lines would have to be long enough to handle the largest element, which would be wasteful for the majority of elements which were smaller. The type of structure described can handle elements of any size at the cost of a flag and some minor additional processing.

Suppose that the circuit has 'settled down' to the state shown in Table 6.5 and that simulation time is reset to zero. The element model includes a delay such that an input change resulting in a *0* to *1* change at the output will take 9 ns and the opposite change 5 ns (t_{pLH}, t_{pHL} respectively). At time *0* the test harness architecture causes input A to change from *0* to *1*. Time units are taken as 1 ns.

The fan-out index of A is known by the primary input controller to be FOM1. The fan-out memory (Table 6.6) is accessed at line FOM1. The state table address is read as line ST5 and the value is read from there as *0* (Table 6.8). Comparison of the new value of A, *1*, with the value from the state table shows the two to be different. The state table value is updated. The network memory reference in line FOM1 of the fan-out memory is to line $NM(1+2) = NM3$. The value in line NM3 of the network memory is updated and all the data for G1, including its type, are sent to the evaluator. This is lines NM1 to NM5. As the second input (B) is still *0*, the predicted output is *1*. This is not a change, but, as indicated in Section 6.3.2, it must still be sent to the event memory. The delay to a *1* is 9 ns, so an entry is made in the time wheel at time 9, Table 6.7. The time wheel pointer is set to EM0 and location EM0 of the event memory is set to contain the value *1* and the fan-out index of FOM5, read from line 1 of G1 in Table 6.5. This index is the line in the fan-out memory that will be referenced to get the fan-outs later. Table 6.9 shows the history of the free list. The event memory address used is the address at the front of the list. This table should be consulted both when building event memory lists and when processing the events of Table 6.7.

The flag in line FOM1 of the fan-out memory, which refers to input A, is *1*, indicating that another gate input is driven by A. A further line is read. This points to line $NM(6+2) = NM8$ in the network memory. The value on that line is updated, and the data of G2 are sent to the evaluator. Both inputs are *1*, so the output is predicted to change to *0* at time 5. The next free memory location in the event memory is at address EM1. The value in this location is set to *0* and the fan-out index to FOM8, Table 6.7, found from line NM6 of the network memory. The time wheel pointer is set to *1*. Notice that this event is 'earlier' than the previous one in the time wheel.

Table 6.7. *Time wheel and event memory for example*

Time	Time wheel ptr[a]	val	foi	ptr	val	foi	ptr	val	foi	ptr
0	null									
5	EM1	0	FOM 8	null						
9	EM0	1	FOM 5	null						
14	EM1	1	FOM12	null						
100	null									
105	EM1	0	FOM 5	null						
	EM0	0	FOM10	EM1	0	FOM 5	null			
114	EM0	1	FOM12	null						
	EM1	1	FOM 8	EM0	1	FOM12	null			
	EM2	1	FOM10	EM1	1	FOM 8	EM0	1	FOM12	null
119	EM1	0	FOM12	null						
123	EM2	1	FOM12	null[b]						
200	null									
209	EM1	1	FOM 5	null						
	EM0	1	FOM 8	EM1	1	FOM 5	null			
214	EM0	0	FOM10	null						
218	EM1	1	FOM 8	null						
223	EM0	1	FOM12	null						

[a]Time wheel pointer is the event memory address.

[b]This event must be removed by some means or there will be an error – see later. Event memory location 2 is presumed to be recovered after location 1 when processing time 114.

As the flag in line 2 of the fan-out memory is *0*, there are no further fan-outs of signal A. Time now advances to find the next event, which is at time 5 in the time wheel, Table 6.7. It should be appreciated that this circuit is very small and hence the time wheel is very sparsely used. Experiments on large circuits suggest that, if the time steps are chosen not too small in relation to the gate delays, then most time steps contain some events once the initial build up has taken place. Indeed, this example illustrates the need to match the time step to the smallest gate delays. With the figures given (5 and 9), a better match would not be possible.

At time 5 the time wheel pointer is EM1. Address EM1 of the event memory has a fan-out index of FOM8. Reading line FOM8 of the fan-out memory (Table 6.6) finds the state table line ST2. Considering Table 6.8

Table 6.8. *State table for example*

ST	T	T	5	9	14	100	105	114	119	200	209	214	218	223
addr	<0	=0												
ST1	1			(1)			0			1				
ST2	1		0					1			(1)		(1)	
ST3	1						0	1				0		
ST4	0				1			(1)	0					1
ST5	0	1							0					
ST6	0					1								

line ST2, it is found that the value is *1* prior to time 5. The value from the event memory is *0*, so there is a change. The value in line ST2 of the state table is updated as shown in Table 6.8. The fan-out memory line FOM8 points to line NM(6 + 1) in the network memory. This is the output of G2, D, which is also updated. As this is an output, an evaluation is not required. Line EM1 is returned to the free list, Table 6.9.

Line FOM8 of the fan-out memory has its flag set, so a further line is read. This points to line NM(16 + 2) in the network memory, an input of G4. This gate must be evaluated. The data are read from lines NM16 to NM20 of the network memory. The inputs have the values *1* and *0*. The evaluation predicts that Z will change to *1* in 9 ns, at time $5 + 9 = 14$.

Event memory location EM1 was recovered when reading the event at time 5, so this is the next available location, see Table 6.9. It is written with the value *1* and the fan-out index FOM12 from line NM16 of the network memory (G4). The time wheel pointer at time 14 is set to EM1, the event memory location used.

The flag at line FOM9 of the fan-out memory is *0*, so there are no further fan-outs for this output, D, to be considered. Fig. 6.15 shows the waveforms.

Time now advances to 9, the next slot with an event (Table 6.7). The time wheel pointer is EM0 and event memory address EM0 is read. Address EM0 is returned to the free list. Note that the free list is now EM0,2,3 ... Event memory location EM0 contained the fan-out index FOM5 and value *1*. Fan-out memory location FOM5 refers to state table location ST1, which contains the value *1*. This is the same as that from the event memory so no further processing is required. In Table 6.8, a bracketed value is shown to indicate that a comparison took place but there was no change.

The next event to be found is at time 14. The fan-out index is FOM12 and line FOM12 of the fan-out memory points to line ST4 in the state table.

Table 6.9. *History of the free list*

	Comment	Free list
Time 0		0, 1, 2, 3, 4 …
	G1 prediction at time 9	1, 2, 3, 4 …
	G2 prediction at time 5	2, 3, 4 …
Time 5		
	recover EM1	1, 2, 3, 4 …
	G4 prediction at time 14	2, 3, 4 …
Time 9		
	recover EM0 – no change	0, 2, 3, 4 …
Time 14		
	recover EM1 – end	1, 0, 2, 3, 4 …
Time 100		
	G1 prediction at time 105	0, 2, 3, 4 …
	G3 prediction at time 105	2, 3, 4 …
Time 105		
	recover EM0	0, 2, 3, 4 …
	G4 prediction at time 114	2, 3, 4 …
	recover EM1	1, 2, 3, 4 …
	G2 prediction at time 114	2, 3, 4 …
	G3 prediction at time 114	3, 4 …
Time 114		
	recover EM2	2, 3, 4 …
	G4 prediction at time 123	3, 4 …
	recover EM1	1, 3, 4 …
	G4 prediction at time 119	3, 4 …
	cancel event at 123	2,3,4 …
	recover EM0 – Z	0, 2, 3, 4 …
Time 119		
	recover EM1 – end	1, 0, 2, 3, 4 …
Time 200		
	G1 prediction at time 209	0, 2, 3, 4 …
	G2 prediction at time 209	2, 3, …
Time 209		
	recover EM0 – no change	0, 2, 3, 4 …
	recover EM1	1, 0, 2, 3, 4 …
	G2 prediction at time 218	0, 2, 3, 4 …
	G3 prediction at time 214	2, 3, 4 …
Time 214		
	recover EM0	0, 2, 3, 4 …
	G4 prediction at time 223	2, 3, 4 …
Time 218		
	recover EM1 – no change	1, 2, 3, 4 …
Time 223		
	recover EM0 – end	0, 1, 2, 3, 4 …

Note: it is by chance that the free list ends in numerical order. It would not have been so if the simulation had ended at time 14 or 119.

That contains the value *0*, and the value from the event memory is *1*. This is a change, so the state table value is updated, as is the value in line NM(16 + 1) of the network memory (G4, Z). There are no fan-outs of this signal and there are no more events in the time wheel, so there is no further activity as a result of this change of A. The free list now reads EM1,0,2,3 ...

This would be the end of the simulation, but the primary input controller applies a new change at time 100. This is B going to *1*. Tables 6.7 and 6.8 and Fig. 6.15 give the details of what happens, as well as for a further change of A later on. These will not be described in detail. There are, however, several points of interest.

(i)   The change of B results in output predictions for both G1 (C) and G3 (E), and for the same time, 105. The first of these is for G1 (C) (line NM3 of the network memory). The address at the head of the free list is EM1, and this is set to the time wheel pointer. Value *0* and fan-out index FOM5 are placed in event memory location 1 (Table 6.7, *T* = 105 first line). Address EM0 is now at the head of the free list.

The second prediction is made and placed in location EM0 in the event memory. The time wheel pointer is copied to the pointer of this location, and EM0 placed in the time wheel pointer (Table 6.7, *T* = 105 second line). The prediction of E for G3 is again *0* and the fan-out index is FOM10 from network memory line NM11.

When time moves to 105 these two predictions result in three more predictions, all for time 114. Table 6.7 shows how these are handled.

Fig. 6.15. Waveforms for example circuit and inputs.

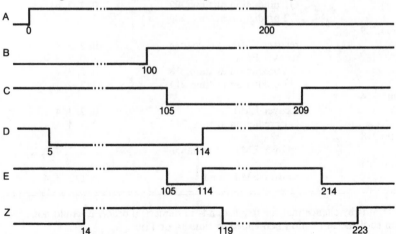

(ii) Consideration of the potential changes at time 114 illustrates another problem. The first indicates a change of E to *1*. As D is still *0* (data not yet read) a prediction that Z will change to *1* at time 123 is made. However, on reading the D change it is found that Z is predicted to go to *0* at time 119. Consideration of the circuit shows that the change predicted for time 123 must not be allowed to happen. The mechanisms described so far would cause a wrong output. It is assumed that the event at time 123 is deleted after the second event is read on processing $T = 114$, and so the free list at time 120 reads EM1,0,2,3...

This problem is a consequence of there being two (or more) changes of the input of a logical element at the same time, and there being an evaluation of the element G4 for each change. There are two solutions, *both* of which are required. For this specific case it is sufficient if it can be arranged to evaluate each element not more than once in each time slot. This is described further in Section 6.5.1. Note that yet another table is required.

The second solution is to remove the event at time 123. This in turn requires knowledge that the event has been placed, followed by a search of all event memory lists which might contain the false prediction. Even though the number of lists to be searched is restricted, and may be limited to one, with large circuits such a search could take a long time, since linked list searches are essentially serial in nature. In software on a single processor, it may not be serious. Where an attempt is made to use parallel processors, it is potentially disastrous. However, the use of unequal rising and falling delays, as here, makes it necessary to have some such mechanism. Chapter 7 on modelling will discuss this in more detail.

There are two further points. Fig. 6.15 shows a short pulse on E. This would really occur, and care must be taken in the design process to avoid such a waveform being used as a clock signal, for example. Referring back to Fig. 6.2, it is seen that the 'logic only' LCC simulator does not predict this.

Secondly, for the input change at $T = 100$, all the outputs have changed by time 119. It might be assumed at this point that the longest delay through the circuit is 19 ns. As Fig. 6.15 shows, changing A (or B) to *0* at 200 ns will result in a 23 ns delay for the circuit to settle. This points to the need for great care in the design of test programs. In fact, no real test program could be expected to analyse all these **race** conditions, so there is a need for timing verification. A glance at the circuit in Fig. 6.14 suggests that the longest delay is three gates, and for the largest value this will be

$(2*9+5)=23$ ns, rather than $(2*5+9)=19$ ns. This is a crude timing verification but real circumstances are rarely as simple and it may well be pessimistic in particular circumstances. The aim of a timing verifier is to perform a full analysis of the circuit independent of the input patterns and this is described more fully in Chapter 8.

6.5   Some refinements

6.5.1   Affected component list and memory

As described so far, each input change is sent in turn to an evaluation routine and the effect of the change determined. If two inputs of the same logical element change at the same time step, then the routine will be entered for each change.

- Such a case will require two predictions to be made. A suitable model will do this (see below).
- On the other hand, it may well result in a situation where contradictory predictions are made, as in the example at time 114. Once a prediction reaches the event memory it is costly to remove it.
- In the case of a flip-flop or register where the clock and data inputs change together and the clock change is evaluated first, the wrong prediction is made (this assumes zero set-up time. It will be seen in the discussion of modelling that 'data' is a pseudo-signal, and that this condition is, indeed, possible).
- In the case of complex elements, evaluation routines may be quite long, and running them adds significantly to the time to simulate. Entering the routine several times in one time step is most undesirable.

What is required is a mechanism to scan all changes for a given time step before any of them causes an evaluation. Fig. 6.16 is a revised version of Fig. 6.6. Before any evaluations are performed, all events are extracted from the event memory and all fan-outs followed, so as to build a list of **affected** elements. The changes are assembled in order of the affected elements – that is, for example, in the order in which the elements occur in the network memory. To do this, an **affected components list** (**ACL**) is set up between the fan-out memory and the evaluation routines (Fig. 6.7). This is yet another table of pointers, one for each record in the network memory. The address read from the fan-out memory accesses the affected components list. The ACL pointers point to locations in an **affected components memory** (**ACM**), Fig. 6.17. The affected components memory will hold the value for each

Fig. 6.16. Outline flow of simulator with ACL and ACM (Fig. 6.6 modified).

Time $= T = 0$
Repeat
   While (events for time T in test program)
     place event into main set for time T

Repeat (with set of events for time T)

   extract event
   Repeat
     find a fan-out from fan-out table
     extract data for affected element
     evaluate affected element model
     Repeat place predicted output in set for time $T + \delta t$
     Until no more outputs
   Until no more fan-outs

Until no more events

Increment T

Until (all event sets are empty AND end of test program)
   OR run out of computing time.

Fig. 6.17. Affected components list and memory.

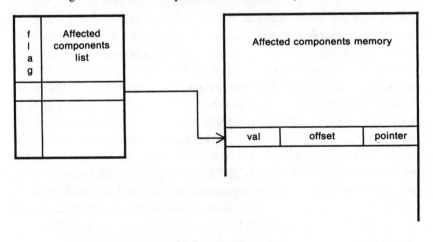

change and the position within the network memory record. It will also have a pointer field. When a change arrives, the ACL is accessed. If it is *null*, the next empty record in the ACM is found and its address placed in the ACL. The value and offset of the ACM record will be set. If the ACL pointer was not *null*, its value placed in the pointer field of the new ACM record. Thus this new change is added to a linked list of changes, all of which refer to inputs or outputs of the same logical element. This structure and procedure is similar to the time wheel and event memory.

When all changes for the current time have been assembled, they are read from the affected components memory. All changes for one element are read by following the linked list, the start of which is the affected component list record. The data is fed to the evaluation routine and the model evaluated. This is repeated for all elements, but if the ACL pointer is *null* there are clearly no changes for that element.

One refinement is necessary. Output changes have to be sent to the network memory as well as input changes (unless the state table is being used for that purpose). If an element has ONLY output changes, an evaluation is unnecessary. To avoid such evaluations, a flag can be set in the affected components list record whenever an input change is received, but not for output changes. This may well be 'set' more than once, which does not matter. It is unset on reading the list.

6.5.2   Time wheel overflows

In any sensibly arranged simulation, the smallest delay of interest will be identical to the time step of the simulator. Under these circumstances, the vast majority of logic delays will be less than the span of the time wheel. This will result from a combination of appropriate time scaling in the simulator environment and the design of the simulator. However, there is always the possibility of longer delays. A common possibility is an instruction in the test program to delay the next primary input drive by 'a long time'. This may result in a special event in the time wheel which will reactivate the test program when this time delay expires. Another example might be in a simulation of a telecommunications system where the response from the line might be milliseconds but the logic element delays are in nanoseconds. Suitable provision must be made for these cases.

One way to handle them is to create a time ordered linked list of events which are scheduled to happen beyond the last time wheel slot $(T+n-1)$. The records in this list will be similar to those in the event memory, but must also include the value of time. Such time ordering is acceptable for this list where it was not for the main set of lists, because it is expected that there will be few events placed here. Finding the right place to put in a new event should not take too long. As simulation time advanced to the point where the event at the head of the list is at time $(T+n-1)$, these events are transferred to the time wheel. Alternatively the time of the event at the head of the overflow list could be compared with current time (T) and taken directly to the fan-out memory along with those from the event memory.

Another possibility might be to allow time to advance until the event at the head of the overflow list is at some other value within the range T to $(T+n-1)$. T is a possibility. The overflow list is then transferred to the time wheel for all events up to $(T+n-1)$.

6.5.3 Wiring delays

The delay between the input of one logical element and the input of those driven by it consists of three parts.

- The basic delay of the logic element. This is a function of the logic element and is built into the model.
- Increased delay due to loading of the circuit. This is a function of the input capacitance and resistance of the driven circuits and the driving capability of the logical element. It is, therefore, network dependent, and cannot be built into the logic model.
- Wiring delay. In some cases this can be incorporated with circuit loading. In others the wire is resistive as well as capacitive (e.g. polysilicon connections on an MOS circuit), and R–C lines behave differently from simple R–C networks. On a PCB with high speed logic, the connections must be treated as L–C transmission lines. This is a function not only of the network, but of its layout. This will not be known when early simulations are run.

The basic gate delay is built into the logic model. The loading effect could be added as a parameter to the output delay of the model. This would be held in the network memory, loaded with the rest of the network, and passed to the model with the rest of the network memory data. One system, from IKOS, uses load capacitance and slew rate constant (in nanoseconds per picofarad) of the signals to calculate the extra delay (Fazakerly 1988). This paper also points out that certain MOS circuits also need input delays added to cater for the effect of slow input rise and fall times on element delay. These can be added as indicated below for the wire delay.

A similar procedure might be used for wire delays. For the reasons just explained, this is added to each input of each driven circuit. The model becomes more complex. Fig. 6.18 shows a two-input AND gate. The delay on output is the basic gate delay enhanced by a second delay to represent the effect of the load elements. These can be combined. The input delays add two pseudo-nodes, each of which needs space in the network memory and on each of which there are events. Thus there are more events in the event memory and a larger state table.

An alternative is to add an extra time wheel and event memory between the fan-out memory and the affected components list (Fig. 6.16). The number of time intervals would be a good deal less than that used by the main time wheel. Since the number of events passing round the simulator loop is stable, the number of events might be expected to be about the same. That is, the increase of 3.5 due to fan-outs is reduced by the state table finding non-events.

The wiring delays are precalculated since they are constant for a particular network. A table of delay per unit length for each type of wire that might be used is held alongside the network compiler. The network then supplies the length of each wire.

The designer of the simulator has to decide whether the cost of the storage required in either case is worth the extra accuracy achievable. This special time wheel would be bypassed most of the time, since the necessary layout data would not be available. It would only be used for a final run.

6.6   Groups of signals

6.6.1   Usefulness and problems

All the descriptions so far have assumed individual signals. There are many places in digital design where it is convenient to treat a group of signals as a unit. Examples might include

- memory address signals,
- data buses in a processor.

Treating such groups as a unit leads to more compact data representations and to a reduction in processing.

Consider the event memory of a signal, Table 6.7. This consists of a value and a fan-out index. The fan-out index is an address and may well be 20 bits

Fig. 6.18. AND gate model with load and wire delays.

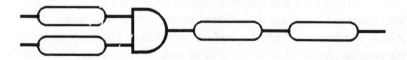

or more. The value in a 46-value system will be at least six bits. Thus an event memory entry will need to be at least 26 bits and probably longer.

Suppose several signals to be combined in a group. The group requires only *one* fan-out index. If four values are packed into a 32-bit word, say, and one additional word is used, the group requires just two words instead of four.

Since the group is treated as a unit, it requires only one fan-out index. As there is only one set of fan-out data per group, the storage requirements of the fan-out memory are also reduced (or conversely a larger system can be simulated with the same resources). Savings can also be achieved in the network memory, but these are relatively minor. Finally, as the group is a single unit, only one event has to pass around the simulator for each group when several (many) signals can be expected to change at the same time. Even though the data for a group is larger than that for an isolated signal, it is much less than for a fully split group. Thus the simulator runs faster.

The primary problem with including groups of signals lies in the region of the state table. It will be recalled that the state table shown in Fig. 6.7 is mainly for convenience of description. Physically the data resides in the network memory. In practice, both multiple output elements and groups would cause problems if the state table were physically separate. These problems are beyond the scope of this text.

For the present purpose, the problem to be solved is how to compare the signal values from the event memory with those in the state table. It will be noted that all that needs to be known is their equality or not. Thus it is possible to compare several values at the same time. Suppose, for example, the group values are packed four to a 32-bit word. A 32-bit comparator giving equality (or not) can compare four values at once. Suppose that there are 24 values in the group. The first four compare as being equal. An inequality may still occur, so the next four values are compared. If they compare as being not equal then the comparison process can be ended immediately – the new and old sets of values of the group are different.

6.6.2 User-defined values

The reader may well wonder why a word such as 'bus' has not been used for a group. The reason is that the representation of a bus is just one use of this mechanism. It is, in practice, a very general facility.

As an example, suppose a floating point arithmetic unit is to be simulated. This may well perform a series of functions on up to 80 bits (double extended format as defined in the IEEE standard). A floating point number consists of three portions. In a 64-bit (double) format these would have one, 11 and 53 bits (total 65, one is not stored, but is needed in arithmetic). If each bit is represented in, say, five-value form, the simulation will be most complex. Indeed, at some point this may well be necessary. However, such complexity is only necessary when doing detailed design of the arithmetic unit itself. In the early stages of system design when the detail is unknown, and in the later stages of system design when the unit has been tested sufficiently to be accepted as 'good,' a two-value representation of signals may be acceptable. In these cases, each bit of the floating point number can be represented by one bit of a group. If the 'normal' group values are of eight bits, then 10 such values can represent the 80-bit number or nine the '64'-bit one. The point at issue is that the *meaning* of what is in the value part of *any* signal or group is irrelevant in the various memories of the simulator, so long as the *evaluator* can correctly interpret that meaning. *Only* the evaluator needs to know the meaning. To the rest of the simulator it is merely a collection of bits. VHDL integers etc. recognise this fact.

The value to the simulator of the floating point unit is that once the evaluator has assembled the various values into 64-bit, 80-bit etc. sections, the floating point simulation can be done by the floating point hardware of the engine running the simulation – a hardware modeller without special hardware (see Section 7.12). Relative to any other method of evaluation, this is *very* fast, even allowing for assembling and disassembling the data.

One point of warning. When using the instructions of the host machine in the evaluation, it is important to be sure what those instructions do. For example, some floating point units are specified as working with numbers in standard format. That may be very different indeed from giving results which conform to the standard. Furthermore, many machine and chip manufac-turers claim conformance to the standard, but when reading the small print there are exceptions. In some cases one can only suspect exceptions. Similar comments apply in other areas. The point is *TAKE CARE*!

6.6.3   Group splitting

Suppose a data word is read from a memory as a group of 32 signals and placed in a processor instruction register. The processor now wishes to treat eight bits as instruction and 24 bits as an address of data. There is a need to be able to **split** a group into smaller groups, possibly even individual bits.

Suppose that some data has been received by a piece of logic as a series of bytes from an external device. These are to be combined in fours to form 32-bit words. There is a need to **combine** two or more groups to form a larger group.

Group splitting and group combination are both characteristics of the *network* and not of the logical elements. For example, consider the instruction register example. It is required to split the 32-bit group into subgroups of eight and 24 bits. The register model is a single element with a 32-bit group of data at the input and a 32-bit group of data at the output, together with clock and control signals. The same register model is used for other 32-bit registers – scratchpads, integers etc. Furthermore, other registers may require other sub-divisions.

When the register has an input change, the data is sent to the evaluator and the output predictions determined. These are then sent to the event memory with a fan-out index in the usual way. There is a bit to indicate that this is a group and some means of identifying the size of the group. The full group must *always* be sent to the event memory because the outputs in the state table and the network memory must be updated at the proper time.

The fact of there being subgroups is due to some of the signals being fanned out to places to which other signals do not go. Fan-out information is needed for each subgroup. This points to a method of splitting. A new routine is placed in the simulator between the evaluator and the event memory (Fig. 6.7). When the network is compiled, this unit is loaded with the label of any group which has to be split. The label can be the fan-out index, since this is unique. Associated with this is a procedure indicating the split(s) required.

When the 32-bit register prediction reaches the group split unit, the unit notices that it is a group rather than a single signal. It consults its table of groups to be split and finds that this is such a group. The full group is sent forward to the event memory as indicated above.

The group split unit now calls the splitting procedure for this group. For the example quoted two new groups are formed. These have eight and 24 bits respectively. The network compiler has assigned a fan-out index for each. It is attached to the value and the time_ahead data and sent to the event memory in the usual way.

Fig. 6.19 shows a possible scenario. R1 has fan-out index 53, and line FOM53 in the fan-out memory will point to the network memory representation of R1 to set the outputs. FOM54 will point to R2 as a load register. The network compiler will have given the first subgroup to fan-out index 55. The subgroup will be specially marked as having no 'output' to set, but it must still have a state table entry and comparison. FOM55 will give the network memory address of R3 and FOM56 that of R4. Next, FOM57 will point to the network memory representation of mem_1 and FOM58 will point to mem_2. The flags will be set for lines FOM53, 55 and 57 and unset for lines FOM54, 56 and 58. Note that it is quite possible for one subgroup to compare as being 'equal' in the state table, and the other to compare as being unequal.

6.6.4   Group combination

Group combination is essentially trivial. Each subgroup has fan-out data in the fan-out memory consisting of a network memory line for the start of the load element and an offset. Suppose that a combined group starts at NM(1234 + 5). It consists of an 8-bit group, two single bits and a 20-bit subgroup. Each subgroup, including the single bits, will have address NM1234 in the fan-out memory. The 8-bit subgroup has offset 5. It will occupy network memory location NM1239 to NM1246. The two single bits will have offsets 13 and 14 and so will occupy locations NM1247 and NM1248. The 20-bit subgroup has offset 15 and will occupy locations from NM1249 onwards. All these offsets are computed by the network compiler at compile time.

It will be realised from this that the subgroups must be contiguous sets of bits. If a subgroup is to be formed from alternate bits of a group, then the group must be split to its constituent bits and then recombined into the required new 'subgroup.' All this is invisible to the user, of course.

This chapter has limited itself to the basic compiled code and event driven algorithms. A number of alternatives have appeared in the literature. None has so far caught on. The references mention a few of these.

The major times involved in simulation are

- calling and evaluating models,
- extracting data from the timewheel, following fan-out linkages and forming the affected components list,
- searching for erroneous data.

Fig. 6.19. Group splitting.

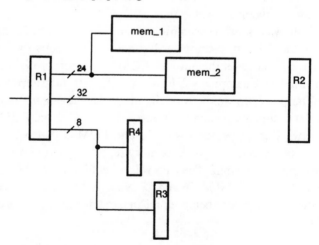

The modelling methods to be described in the next chapter will emphasise the avoidance of the last point.

Switch level simulators require different modelling approaches but many of the simulators described in the literature use or adapt the techniques described here. Again, the references give some indication of what this author feels is the most useful work.

7

Models and model design

7.1   Some simple models

The previous sections on simulation algorithms and simulators have assumed that when the input to a logical element changed, some computation was performed on the input values and an output value (or output values) was predicted. This output prediction might or might not be a change from the previous value(s) of that/those output(s). The process which transforms the element inputs to output(s) is called a **model**.

Models are not something new. Consider Fig. 7.1, in which a resistance, R, is connected across a battery, B. From Ohm's law write $I = V/R$; but think again for a moment. It has been assumed that V is a constant. That is not true of any real battery. In this calculation of current an **approximate model** of the battery has been used. It would be more accurate if a small resistor were put in series with it.

The assumption that R is a constant is also an approximation. In most

Fig. 7.1. Model of a simple electrical circuit.

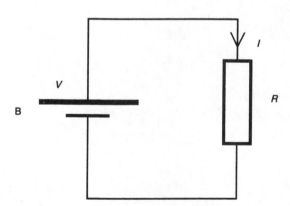

Table 7.1. *Two-input AND gate*

A	B	C
0	0	0
0	1	0
1	0	0
1	1	1

cases the value of R will vary with temperature. When current flows I^2R watts are dissipated as heat. The value of R will alter, causing an adjustment of current and so on. The value of R is a function of I. For some purposes this might be important, so a more complex *model* of R is needed. Notice that the power is not expressed as VI, since the voltage across R is also a function of I, especially for the 'real' battery.

For the present purposes, the simplest model for a digital device is a boolean equation or a table of values. For example, for a two-input AND gate, given inputs A and B, the output, C, is given by C = A & B, or by using Table 7.1. If one or both of the inputs change, C may assume a new value δt ns later. In a model for a 32-input adder, one might express the inputs and outputs as three bit-vectors of 32-bits. The model could still use a table, but it would be enormous. It is more 'efficient' to express the input and output as 32-bit integers and to use the adder of the CPU to execute the model.

At this point it is as well to be clear about certain matters.

- Modelling involves taking a set of element inputs, *predicting* changes in the output(s), and can involve *internal variables* which may be set immediately or predicted.
- For digital simulation there are only two defined logic values per bit, which in this text are denoted by *0* and *1*. The electrical interpretation of these will vary from circuit to circuit. Indeed, there is no need for them to be electrical at all. They might be light intensity or pneumatic pressure. Furthermore, there is no need to associate *0* with a lower voltage than *1* – it could be the opposite.
- For the purposes of digital simulation, all changes of value take place instantaneously. In other words the rise time and fall time of all signals is *zero*. This will continue to apply when additional values are defined later.

It is understood, of course, that real signals do not have zero rise and fall times. Should such real values affect the circuit delays then the model of the elements must be modified to include rise time effects. This will also be considered later.

- The simulator itself should have *no* knowledge of the circuit or signal representation being used.
- For a (simple) compiled code simulator, the model will be limited to boolean equations or a table. Time is not included. Most of the 'problems' of model design relate to timing, so models for an LCC simulator will be relatively simple. The YSE uses a table for four-input one-output devices. Such models can be used with an event driven simulator but that would be to defeat the primary advantage of such a simulator – the ability to reflect real timing. Much of what follows, therefore, presumes an event driven simulator. It is not too difficult to find the exceptions. Some of what is said will also be applicable to timing verifiers discussed in Chapter 8.

In many texts the authors write of **functional, behavioural** and **structural** models. There is no universally agreed definition of what these are. The best the author has seen is that of Abramovici (1990) which gives the following.

> *Functional model* describes the logic function only and no timing. These are mainly of use in LCC simulators.
> *Behavioural model* describes logic *and* timing. These are used as basic elements in event driven simulators. They need not be simple elements.
> *Structural model* describes a box as an interconnected set of smaller boxes. The 'structure' is in the connections. A schematic diagram (a box) is a structural model – an interconnection of smaller boxes such as gates, ALUs, shift registers etc., or even other diagrams. At the lowest level, the boxes have behavioural or functional models.

Most of the rest of this chapter will discuss the construction of behavioural models of simple logical elements.

7.2   Delays

Most simulators define at least two types of delay. In a simple high conductivity (copper or aluminium) connection the signal at the output end of the wire reproduces exactly the shape at the input to the line. This is known as **transport delay**, and is illustrated by OUT(T) in Fig. 7.2.

When a signal passes through a gate, it takes a certain amount of energy to cause the output to change state. If the input is narrow, the amount of energy supplied is insufficient for this purpose. It is said that the input pulse cannot overcome the *inertia* of the gate. An **inertial delay** is defined as one which exactly reproduces the input shape at the output *provided* that the time between two edges of the input in opposite directions is more than some minimum, t_i. If the time between the edges is less than t_i, no output change takes place. This is illustrated in Fig. 7.2 by OUT(I). In many simulators, t_i is set to the gate delay, t_d. The value of t_i will be different for the two edges in the general case.

The following description will presume that all delays are transport delays, since this is all that can be implemented directly. However, *in most simulators the default delay is an inertial delay*. Section 7.3.4 will describe how an inertial delay can be realised.

7.3 Model of a buffer

7.3.1 Development of the algorithm

A buffer is a simple device which in principle takes one input signal and produces an output signal which is identical to the input. Such a circuit is usually used to take a relatively low powered signal whose rise time would be badly affected by the capacitive loading of many driven circuits and boost the drive power – in other words, a current amplifier. As the logical function is simply 'copy' the timing effects of the model are the sole interest. An inverter can also be regarded as a buffer whose function is 'copy and invert.' In an LCC simulator both a buffer and an inverter would be removed from the network description by preprocessing.

The first departure from the ideal is that any logical element introduces **delay**. Again, an initial approximation is to say that the delay is independent of the direction of output change – *0* to *1* or *1* to *0*. Unfortunately most real circuits do not work that way. Thus the model requires two time parameters. These will be called t_r for a *0* to *1* change and t_f for *1* to *0*. In many places, these are called t_{pLH} and t_{pHL} respectively. Fig.

Fig. 7.2. Transport and inertial delays.

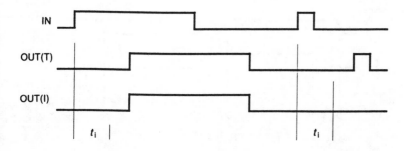

7.3(a) shows the meaning of this. It is again emphasised that t_r implies an *instantaneous* change from *0* to *1* after an appropriate time delay from the input change, and must not be confused with the 'rise time,' t_{rise}, of a real signal changing from *0* to *1* as shown in Fig. 7.3(b).

It will be noticed from Fig. 7.3 that the output wave shape is not the same as the input. It would be if t_r and t_f were equal. However, t_r has been set to 10 ns and t_f to 7 ns. The input is 30 ns at *1* and 30 ns at *0*. The output spends only 27 ns at *1* but 33 ns at *0*. The difference is small, but, continued over several stages, could be important. If there are five such gates in series, the signal would spend 15 ns at *1* but 45 ns at *0*. For clock signals which might be running close to the maximum speed of the clocked devices and which require a minimum width for both *1* and *0* part periods, this is significant.

Consider, now, what happens if IN changes twice close together. In Fig. 7.4 IN goes to *1* at time 0 and back to *0* at time 2. The real circuit would not respond. The simulation predicts an output change to *1* at time 10 and a change to *0* at time 9. Considering the event driven simulator, the event at time 9 is found and there is no output change. At time 10 an event making the output *1* is found. On this basis the simulator gets the wrong answer. The model is too simple. *Note that this effect is due to the difference between t_r and t_f, and not due to inertial delay.*

The solution is to remove the event causing the output to go to *1*. Taking

Fig. 7.3. Delay of a buffer: (a) ideal signals in a simulator; and (b) more realistic waveforms.

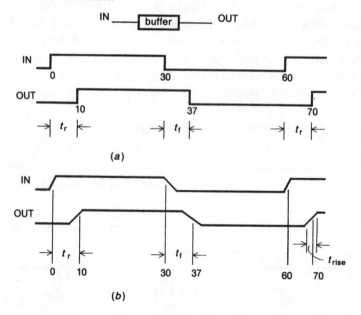

this example, one procedure is that changes of *0* to *1* are scheduled for 10 time units ahead. When a *1* to *0* change occurs on the input, a prediction is placed 7 units ahead and the system then searches for a *0* to *1* prediction for this output in time slots 7 to 9 units ahead of current time. The input change must be at least one unit after the *0* to *1* change; hence it is only necessary to search up to 9 units ahead of here rather than 10. If a *0* to *1* prediction is found, it is removed.

This is quite possible to do. However, the event lists could be quite long and are in random order. As linked lists will be searched serially, the search could take quite a long time.

The computation for this can be reduced. Whenever a *0* to *1* change takes place, the time is recorded by the model in the network memory as an extra value to those described in the last chapter. When this element is next evaluated, a check against this time is made. If it is far enough in the past – 3 time units or more in this example – then the search of the event lists is not necessary. This can be incorporated as a function to check the stability of a signal for a specified length of time backwards from current time.

Consider the input waveform, IN, shown in Fig. 7.5. At time 0, IN changes to *1* and the time is recorded in the network memory. At time 20, IN changes to *0*. A check on the difference shows that $20-0$ is greater than $t_r - t_f = 10 - 7 = 3$. No searching is necessary.

Fig. 7.4. Problem with a spike.

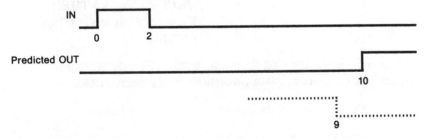

Fig. 7.5. Model of a buffer.

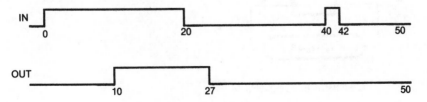

At time 40, IN again changes to *1* and time is recorded in the network memory. At time 42, IN changes back to *0*. $42-40=2<3$, so the event memory must be searched. As it is known when the input changed (time 40) and t_r is 10, it will be sufficient to search the list at time 50 only. The earlier event is removed.

This is satisfactory here. However, for multiple input devices, the delay to an output will vary with which input is changing. In this case, it may be necessary to search several time slots for relevant events.

It should be noted that this problem occurs for only one edge. With a narrow *0*-pulse the *1* to *0* change will cause an output change 7 ns ahead and the *0* to *1* change causes an output change 10 ns ahead. Hence the latter can never 'overtake' the former. These are not incompatible. The output pulse will be wider than the input.

Thus is deduced a very important principle in model and simulator design. *Models should be designed to avoid having to cancel events.*

An alternative to searching lists is to postulate a special 'logic' function, F, and a pseudo-variable, *s*. The logic function has the following definition. 'T' is a time variable held in the network memory representation of the element and *NOW* is the VHDL term for 'current time'.

CASE input_change IS -- from event memory
        WHEN $(1->0)=>$ $F<=0;$ $T<=NOW;$
        WHEN $(0->1)=>$ *s* to event memory at
                                $(NOW+t_r-t_f);$ $T<=NOW;$
        WHEN $s=>$ IF (input$=1$) AND
                                  $(NOW-T\geq t_r-t_f)$ THEN
                                  $F<=1;$ (ELSE do nothing)
                                  ENDIF;

Fig. 7.6 shows the 'logic' of the buffer. The 'delay elements' are indications of where to place predictions in the event memory. t_b is the

Fig. 7.6. Model of the buffer.

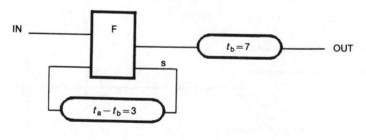

$t_a=\max[t_r, t_f];\ t_b=\min[t_r, t_f]$

smaller of t_a and t_b, in this case being $t_f = 7$. Thus, for an input change 1 to 0 as at time 20 in Fig. 7.5, F gives an immediate change to 0 (see algorithm). This prediction is placed in the event memory list addressed by current time plus 7 in the time wheel, namely 27 as a result of the output delay of the buffer. A similar prediction will take place at time 42, with the prediction placed at time 49.

Consider, now, the 0 to 1 change of IN at time 0 (Fig. 7.5). The function, F, causes an event, s, to be placed in the event memory at time 3. The time, $T = 0$, is recorded in the network memory. Considering only this buffer, the next time of interest is time 3, when s is read from the event memory, set on the input in the network memory and the model evaluated. The value of (current time $- T$) is 3, implying that the input has been stable for 3 time units. It is thus safe to make the change of OUT to 1 in $10 - 3 = 7$ time units, which is, again, the output delay of the buffer. This prediction cannot be cancelled by a subsequent change, since such a change cannot occur before time 4 and requires at least 7 time units to affect the output.

The next event to occur is IN going to 0 at time 20. The output is predicted to change at time 27 and time $T = 20$ is recorded in the network memory.

Next IN goes to 1 at time 40. T is set to 40 and a prediction for s is made for time 43. The change of IN to 0 at time 42 changes T to 42 and predicts an output change for time 49.

At time 43, the event, s, is read and causes evaluation of the model. T is now 42, and $43 - 42 < 3$. Thus F is not set to 1 and no prediction is made.

Finally, the prediction F to 0, is read from the event memory at time 49. Comparing this value with the current value in the state table shows no change, so fan-out will not be affected.

The reader should repeat this exercise with several rapid changes of input – to 1 at 60 and 62 and to 0 at 61, say. This will confirm that the algorithm works for many changes. The output should go to 1 at 72 and at no other time as a result of these input changes. An additional complication arises if a further change to 0 occurred at time 63. In this case, there are two input changes – IN and the pseudo-input s – at time 63. The function, F, must ensure that IN overrides s. As the algorithm is written, the change $1 \rightarrow 0$ is considered first in the CASE statement, so correct operation ensues.

Notice that the model is only stimulated when a *change* comes from the event memory. The event s has no value as such. It is merely a device to ensure that the model routine is re-entered at the appropriate time. A short 0 pulse is described below.

The description indicated that an extra event has to be processed.

However, there is no searching of lists. There is a requirement to record in the network memory the time at which the input changes, and a need for an operation within the model to check that value against current time. This is the price to be paid for the privilege of being allowed to specify the rising and falling delays to be different.

Event cancelling, or its equivalent as just described, is common to most models which include real timing. It would seem sensible to suggest that this routine should be part of the simulation environment, relieving model writers of the need to bother themselves with such detail. However, how does the environment solve the problem? One possible answer is that it inserts a pseudo-model after each element behavioural model in a manner similar to Fig. 7.11. This may take a bit longer than using the combined model of that figure but will make the job of the majority of model writers much easier. The description of buffer models in this chapter can be assumed to be that pseudo-model.

7.3.2   State machine representations

The above procedure can be specified in terms of a state machine as shown in Fig. 7.7. This may seem an unnecessary complication for this model, but will be seen to be important with more complex cases. (Readers for whom model details are not essential might skip this section.)

Fig. 7.7 shows three states:

> Definite_0   (D_0)
> Possible_1   (P_1)
> Definite_1   (D_1)

Suppose that the input has been 0 for a long time. The state is *definitely_0* and will remain so until the input changes. Suppose the input changes to 1

Fig. 7.7. State diagram for a simple buffer.

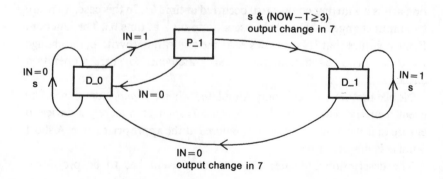

at time 0 as shown in Fig. 7.8, which is a copy of Fig. 7.5 with the states added. It is not known whether the output will change to *1*, so the state changes to *possible_1*. This corresponds to placing the *s* prediction in the event memory.

At time 3, the *s* event returns to the network memory and the model is re-evaluated. The input can now be guaranteed to cause an output change. The state changes to *definitely_1 immediately* but the prediction of output change is made for 7 time units ahead, i.e., time 10.

The state now remains *D_1*, while IN remains *1*. When IN changes to *0*, an output change is guaranteed, so the state changes to *D_0* immediately and a prediction of an output change 7 time units ahead is placed in the event memory. This happens at times 20 and 42 in Fig. 7.5. Notice that there is no *P_0* state, as the change to *0* is always guaranteed.

Now consider the change to *1* at time 40. The prediction *s* is placed in the event memory and the state changes to *P_1*. At time 42, IN changes to *0*. The state must now change to *D_0* immediately. When *s* appears the state is *D_0*, so a prediction of output change to *1* is not made. Such a prediction is only made if the state is *P_1* as at time 3.

As in the suggested example with many rapid changes, the value of IN must be considered before the effect of *s*. Thus, if there are changes to *1* at times 60 and 62 and to *0* at times 61 and 63, then IN and *s* both appear at time 63. The IN forces the state to *D_0* before the effect of *s* is considered. This is part of the function, F. However, suppose that IN goes to *1* again at time 64 (Fig. 7.9). *s* returns at time 65 as a result of the IN change at 62. Although the state is *P_1* this must not be allowed to cause an output change. *A check on T is still needed.*

Notice that the states in the state diagram do not directly imply values for the output, but rather values which will occur some known time after the change to the state.

The short *0* pulse shown in Fig. 7.10 can be examined in a similar manner. The initial state is *D_1*. IN goes to *0* at time 0 and T is recorded as

Fig. 7.8. Copy of Fig. 7.5 showing the states.

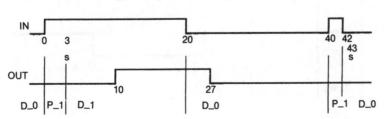

0. The state has gone to *D_0* and a *0* output prediction for time 7 is made. At time 2, IN goes to *1* and T becomes 2. The state becomes *P_1*. Pseudo-variable *s* returns at time 5, and 5−2=3. The state changes immediately to *D_1* and an output prediction of *0* to *1* made for time 12. Notice that the output pulse is wider than the input. Again the states appear to be out of synchronism with the waveforms because changes of state occur immediately while output predictions placed in the event memory take some time to take effect. However, all predictions in the event memory will eventually happen. The purpose of *s* is to avoid incorrect predictions of OUT being made.

In these examples, it is presumed that t_r is greater than t_f, as is generally true for basic TTL gates. If the reverse is true then clearly it will be necessary to reverse all that has been said above with respect to *0*s and *1*s, and the state diagram will contain a state *possible_0* rather than *P_1*.

To implement this model in the event driven simulator, the network memory needs to hold the state variable. Values sent to the event memory include predictions of *0* and *1* as appropriate, together with the SIGNAL, *s*, whose purpose is simply to cause the model to be restarted. The model involves a PROCESS which is evaluated whenever IN or *s* change, so these

Fig. 7.9. Illustrating multiple rapid input changes.

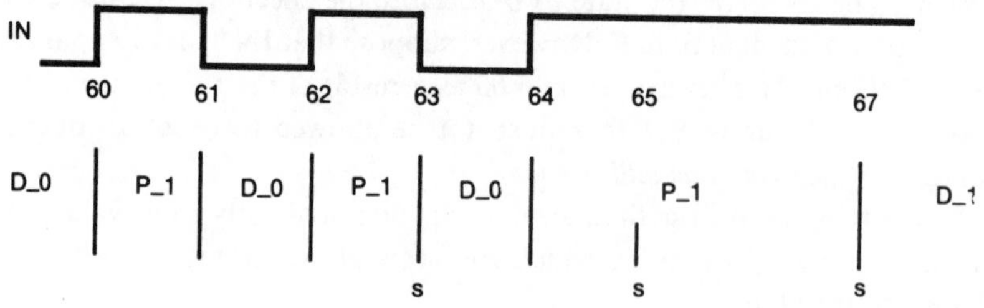

Fig. 7.10. Simulation of a short *0* pulse.

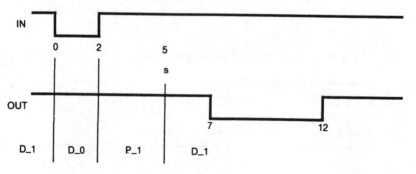

are in the sensitivity list (see Prog. 7.1[1]). The state is a VARIABLE, as it changes immediately. It is held in the network memory. OUT and s are SIGNALs, and are predicted via the event memory. The PROCESS does not use s, but s must be set and reset. There may be more than one prediction of s; a change to 0 after one delta and a change to 1 after some nanoseconds. The change to 0 will restart the PROCESS but, as IN will not

Prog. 7.1. Suggested implementation of Fig. 7.7 in VHDL.

```
TYPE state_1 IS (D_0, P_1, D_1);     -- TYPE definition inside a
suitable PACKAGE
....
PROCESS (IN, s)          -- activate by an input change.
    VARIABLE state : state_1;
    VARIABLE T : TIME := 0;
....
    CASE state IS
        WHEN D_0 = >
            CASE IN IS
                WHEN '1' = >
                    state := P_1;    T := NOW;
                    s < = TRANSPORT '1' AFTER t_r - t_f;
                WHEN OTHERS = > s < = '0';
            END CASE;
        WHEN P_1 = >
            CASE IN IS
                WHEN '0' = >
                    state := D_0;    T := NOW;
                    s < = '0';
                WHEN '1' = >
                    IF ( (NOW - T) ≥ (t_r - t_f) ) THEN
                        state := D_1;    s < = '0';
                        OUT < = TRANSPORT '1' AFTER t_f;
                    ENDIF;
            END CASE;
        WHEN D_1 = >
            CASE IN IS
                WHEN '0' = >
                    state := D_0;    T := NOW;
                    OUT < = TRANSPORT '0' AFTER t_f;
                WHEN OTHERS = > s < = '0';
            END CASE;
    END CASE;
END CASE;
```

[1] This model is designed to run on a simulator with no preconceptions about how TRANSPORT and INERTIAL delays work. It will not run as shown on a commercial software simulator where such an understanding is built in. See Appendix.

have changed, it will do nothing. Selectors in the CASE statement are evaluated in order, so the most likely should be placed first. However, in the case of *P_1*, IN = '0' must take precedence over the option where *s* has restarted the PROCESS with IN = '1'.

7.3.3 Model of a simple gate

The buffer model can be applied to AND and OR gates of any number of inputs. All that is necessary is to evaluate the boolean function prior to the buffer as indicated in Fig. 7.11. When one or more of the inputs IN_A, IN_B or IN_C change, the VHDL VARIABLE, X, is calculated and compared with the old value. If there is no change then the evaluation is complete. If X does change then the evaluation continues with the buffer as described in the previous section. In this case, the AND part of the evaluation has zero delay on both edges, so it is valid to specify X as a VARIABLE and to compare new and old values where that would not have been the case for the element as a whole (i.e. for a VHDL SIGNAL). The other simple functions evaluate in a corresponding manner.

More complex elements cannot usually be calculated as simply as this. In particular a not equivalence gate may be constructed of simpler gates and under particular sequences of input changes may produce an output 'spike.' In other cases, special circuit techniques such as current steering may allow the not equivalence to use the same approach as the simple gates here. The model designer has to be aware of the underlying technology to make this decision. *The reader is warned that the logic diagrams shown in manufacturer's data books are not necessarily a good guide to the circuits.* For example the six-gate logic of the TTL 7474 may be *logically* correct but does not represent the actual circuit which uses current steering.

Fig. 7.11. Three-input AND gate model.

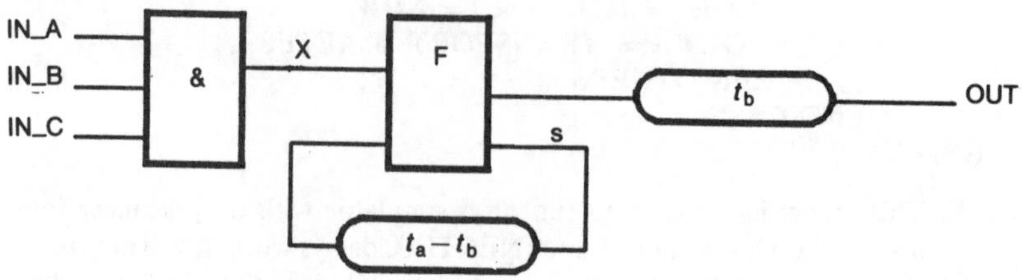

$t_a = \max[t_r, t_f]; \; t_b = \min[t_r, t_f]$

7.4   Inertial delay

(Readers for whom model details are not essential might skip this section.)

7.4.1   Equal rise and fall delays

The model so far described uses transport delays but enables predictions that would be overtaken by later events to be avoided. An inertial delay model enables pulses to be removed when the pulse width is less than some given time, usually the gate delay. It also applies when $t_r = t_f$. Clearly the previous discussion has affinities with this. This section considers the implementation of an inertial delay.

Consider, first, a buffer with $t_r = t_f = t_d$. It is required to remove any input pulse whose width is less than t_d. Fig. 7.12 shows some typical waveforms. The narrow pulses of width less than $t_d(= 10)$ do not appear at the output.

Fig. 7.13 is a possible state diagram. Because it is desired to remove both positive and negative going pulses, four states are used. These are designated Inertial_Possible_0 (IP_0) and IP_1. A pseudo-variable, p, is used in a manner similar to s in the previous discussion.

Consider Fig. 7.12. IN is at 0 and the state is D_0. IN becomes 1, the state

Fig. 7.12. Example of inertial delay.

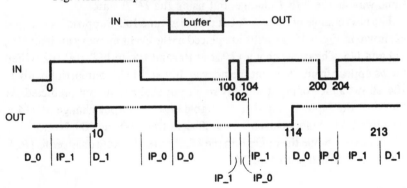

Fig. 7.13. State machine for buffer with inertial delay. $t_r = t_f$.

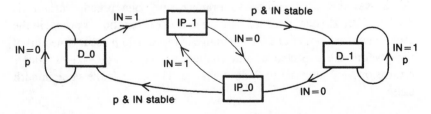

changes to IP_1, time, T, is recorded in the network memory and p is placed in the event memory for time $t_d - 1$ (the reason for the '-1' will appear below). The next relevant event is that p re-activates the model. A check of time against T shows that IN has been stable for $t_d - 1$. By implication, IN cannot change for one more time unit and hence will be stable for t_d. The state changes to D_1 and the output will change in the next time slot.

The reason for placing p at $t_d - 1$ ahead of the input change is related to the use of an affected components list. There can be no guarantee as to the order in which logical elements are processed. If a zero delay is permitted, then an element may be evaluated more than once per time slot since the load elements will have to be re-inserted in the ACL. This is highly undesirable. The procedure described is perfectly valid and avoids the use of zero delays.

An alternative to zero delays and the above is provided in VHDL. Where such zero delays are used, the software automatically provides one or more time steps which are shorter than the simulation time intervals. These are invisible to the user, but ensure that things happen in the correct order. They are known in VHDL as **deltas**. Deltas are used in a number of places, especially where sequential instructions are used to imply an order in which things must be done without changing simulation time.

Returning to the inertial delays, the change 1 to 0 on IN is treated in the same way as the 0 to 1 change, but using the IP_0 state.

The next change of IN is to 1. Let this be at time 100. Suppose IN changes as shown in Fig. 7.12. p events are placed in the event memory at times 109, 111 and 113. The state becomes IP_1 at 100 and 104 and IP_0 at 102. When the ps appear from the event memory at 109 and 111, comparison with T, 104, shows a lack of input stability so output changes are not predicted. At time 113 the input is found to be stable and an output change at 114 is scheduled. The reader can check the effect at time 200 and 209. Notice that, if IN goes to 1 while the buffer is in the IP_0 state, the state changes to IP_1. This is contrasted with Fig. 7.7.

7.4.2 Unequal rise and fall delays

Returning, now, to the case of unequal rising and falling delays, the state machine appears as shown in Fig. 7.14. In this diagram s is the pseudo-variable used earlier. In practice only one pseudo-variable is needed. They are distinguished here for ease of explanation. s occurs in the event memory $t_r - t_f$ after a 0 to 1 change of input, and p occurs one unit of time before the expected output change, namely $t_r - (t_r - t_f) - 1$ after s returns or $t_f - 1$ after a 1 to 0 input change. The delay is the same in both cases.

Consider the input waveform shown in Fig. 7.15 $t_r = 10$ and $t_f = 7$ as before but the inertial delay is equal to the gate delay, i.e. different on the two edges. The first 1 is greater than 10 units long and so will lead to an output change. The following 0 is 6 units long, less than t_f, and so will be ignored. The next 0 pulse is 12 units long, so the output would be expected to go to 0 at $21 + 7 = 28$. The following high pulse is 5 units long and is ignored. The diagram indicates the times at which s and p are found in the event memory and the state machine states.

Time 0: IN goes to 1 and s is placed in the event memory at time 3. The state becomes P_1.

Time 3: the input is stable, so an output can be expected. The state becomes IP_1 and p is placed in the event memory at time 9.

Time 9: the input is still stable so the state becomes D_1 and the output changes at time 10.

Fig. 7.14. State diagram for buffer with inertial delay and unequal t_r and t_f.

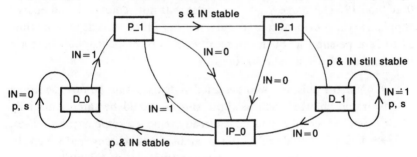

Fig. 7.15. Inertial delays, $t_r \neq t_f$.

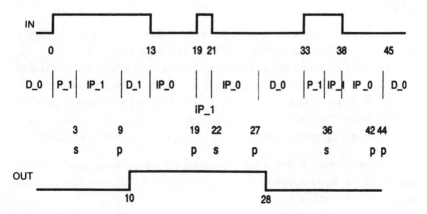

Time 13:   IN changes to *0* and the state becomes *IP_0*. The event memory has *p* at time 19.

Time 19:   IN becomes *1*. The value of IN is considered before *p* so the input is not stable and the state changes to *IP_1*. *p* is ignored. *s* is placed in the event memory at time 22.

Time 21:   IN goes to *0* and the state returns to *IP_0*. A *p* event is placed in the event memory at time 27.

Time 22:   *s* returns and is ignored since the state is *IP_0*.

Time 27:   the *p* event with the state *IP_0* finds the input stable (for 6 + 1) so the output changes at time 28. State changes to *D_0*.

The reader can develop the rest of the waveform, and should try other examples. Output waveforms can be developed heuristically as above to confirm the correct evaluation of the state diagram.

The example shows up a weakness in the model. Considering the physical processes in the gate, it is clear that the output must be close to changing to *0* at time 19. The narrow positive pulse will not remove all the energy supplied in the previous six time units, so a full 7 time units delay from time 21 to 28 is pessimistic. Such complexity could be modelled, but there are two reasons why it should not be.

- Narrow pulses in well designed systems should be rare. Multiple rapid changes such as those shown should be rarer, especially where inertial delays have already removed some narrow pulses.
- There is a trade off between simulation accuracy and speed. In view of the previous statement, is such cost justifiable?

Using inertial delays removes narrow pulses. Using transport delays shows up many pulses which will not occur in practice. There are two dangers.

- Removing spikes may be optimistic and remove some which occur in real hardware with disastrous consequences.
- Leaving many spikes to be reported may result in such a mass of data that the user will not read it. Clearly there is a case for some intelligent system for knowing which spikes are important when using transport delays. With inertial delays, this writer would prefer an inertia less than the gate delays to avoid over optimism.

In the remainder of this book, transport delays will be assumed as that makes descriptions simpler. The above account indicates that inertial delays can be implemented, but at a cost.

7.5   A three-value model

The values of t_r and t_f can be selected to be typical values for the element or could be maximum values. However, if there is feedback in a circuit, it may be important to know the minimum time through the network, or to compare the maximum time through one path and the minimum through another to detect possible short pulses which could cause misoperation, especially with synchronous circuits. Determining which paths should use which delays is very difficult. An alternative is to mark a signal as of uncertain value at some times and to ensure that the signal value is not uncertain at critical points in time.

The uncertain value is commonly designated as X (or sometimes U). A circuit element of a given type will have a minimum value of rise delay of t_{rmin} and a maximum value of t_{rmax}. Consider a buffer with the input signal changing from 0 to 1 at time 0. At time t_{rmin}, the output changes to X and remains there until t_{rmax} when it changes to 1. The value, X, implies that the signal has the value 0 or 1 but it is not certain which. *The statement that the signal changes from 0 to 1 instantaneously remains true.* Thus, when the signal has value X it is *not* in the process of changing, and the time $t_{rmax} - t_{rmin}$ should not be confused with the analogue rise time.

Fig. 7.16 shows an interpretation of the X regions. The values used in the example are

$$t_{rmin} = 3, \ t_{rmax} = 10$$
$$t_{fmin} = 2, \ t_{fmax} = 7$$

The output pulse is somewhere between 32 and 44 time units wide ($42 - 10$; $47 - 3$). The problem with the use of X values is that the X period expands as the signal passes through several stages. For example, Fig. 7.17 shows a signal passing through two buffers (or gates) in series. Suppose that the input changes from 0 to 1. The output of the first stage could change anywhere between 3 and 10 time units. If it changed at 3 units, then the

Fig. 7.16. Interpretation of the X regions.

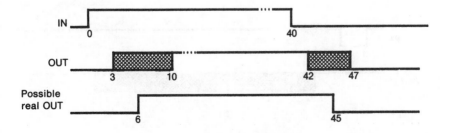

output of the second stage could change as early as 6 units. On the other hand, if both gates were slow, the output of the first would not change until 10 and the second until 20 units.

Even worse, if the output of the first stage drives more than one place then the driven places appear to have a large possible difference between them, whilst in fact they do not because the first stage is the same for both. This is known as **common ambiguity**, and will be discussed further in Chapter 8 (see Fig. 8.2). The important point here is that the resulting waveform is not optimistic.

Consider, now, the diagrams in Fig. 7.18 for a single buffer. The buffer has

$$t_{rmax} > t_{fmax} > t_{rmin} > t_{fmin}$$

which is one of the usual situations. The other main one is where t_r and t_f are interchanged. Values are as for the previous example.

Fig. 7.17. X region spreading.

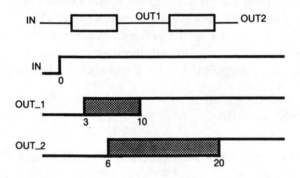

Fig. 7.18. X region with short pulses.

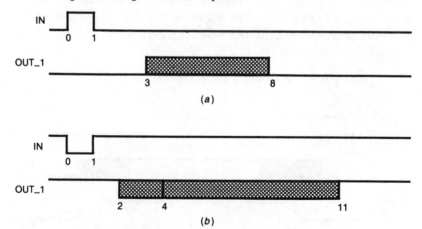

In Fig. 7.18(a) the input goes to *1* so the output *might* go to *1* at time 3. The input returns to *0* at time 1 so the output *could be 0* from time 3. Thus the output would never go to *1*. However, the response to the input going to *0 might not* occur until time $1 + 7 = 8$, giving a 5 unit pulse. The $0 \rightarrow 1$ change at the input could lead to an output going to *1* as late as 9 or 10. However, t_{fmax} ensures that the output is *0* at these times.

Fig. 7.18(b) gives the corresponding situation for a short *0* pulse. The output may go to *0* as early as 2 but might not do so until 7. The output change to *1* could occur as early as 4. Thus, if the output 'went to *0*' at 6 and 'went to *1*' at 5, there would be no output pulse. On the other hand, if the output went to *0* at 2 and to *1* at 11, there would be a 9 time unit pulse. The reader can easily confirm that the conclusions do not alter if $t_r = t_f$. Sequences of waveforms of the form *1 X 1* or *0 X 0* imply a possible short spike. Hence the designer must be careful to ensure that signals such as the asynchronous preset and clear of a flip-flop do not contain such sequences and that signals such as clocks contain them only at non-critical times.

It will be appreciated that any logical change will now require two evaluations; $1 \rightarrow X \rightarrow 0$ or $0 \rightarrow X \rightarrow 1$. Furthermore, if $t_r \neq t_f$, then various rather nasty situations can arise as a result of rapid input changes. The extension to simple gates is as shown in Fig. 7.11.

7.6   A five-value model

Gate models using three values give some information regarding times where signals are uncertain. With short pulses on an input, the spreading of the unknown regions can lead to overlap between possibilities of output changes. There may or may not be possible glitches – short pulses – on the output. Fig. 7.18 is a case where the effects of two edges overlap.

Whether or not a glitch is possible can be determined by defining two extra 'flavours' of unknown. These are designated **rising** and **falling**. *Rising* means that the output was *0* and has become unknown since time is between t_{rmin} and t_{rmax} after the input change. *Falling* has a similar definition. They are illustrated in Fig. 7.19. Yet again it is stressed that *rising* must not be confused with analogue rise time.

Fig. 7.19. Definition of *rising* and *falling*.

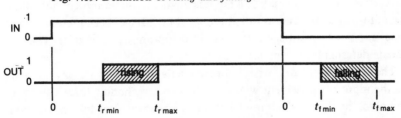

If the input pulse becomes so short that the falling value overlaps the rising (Fig. 7.18(*a*)) then the output is specified as 'unknown,' X, and represents a real possibility of a glitch. All the problems associated with writing the model are again associated with problems of overlapping unknown periods of various flavours. Fig. 7.19 shows a 'clean' input. What happens if the input is *rising* and goes to *falling* or *unknown*? In this case, there are ten possible states in the state machine diagram, namely

Definite_0 (*D_0*)	possible_0 (*P_0*)
Definite_1 (*D_1*)	{possible_1 (*P_1*)}
Definite_falling (*D_F*)	{possible_falling (*P_F*)}
Definite_rising (*D_R*)	possible_rising (*P_R*)
Definite_unknown (*D_X*)	possible_unknown (*P_X*)

If t_r is greater than t_f, as postulated in previous examples, then the state P_1 is subsumed within P_R and D_R. This can be seen from the simple case where an input change from *0* to *1* will lead to the output going to rising and then to *1*. In a sense, these are the same as P_1. Similarly, because t_f is less than t_r, an input change from *1* to *0* can be guaranteed to lead to the falling value. Hence the state P_F is unnecessary. These two unnecessary states are shown in braces above. Corresponding omissions would occur if t_f were greater than t_r.

7.7   Logical combinations and non-logical values

Two problems arise with many valued simulations. Firstly, what is the correct output of a logical function with any given set of inputs, in the knowledge that in the simulations some very odd combinations will certainly occur? The answer is usually set out in a truth table, and refers to an expected output assuming the inputs are not changing too rapidly. Table 7.2 shows the five-value table for AND and OR. For the two-input AND gate, a *0* on one input forces a *0* on the output regardless of the other inputs. If one of the inputs is *X* and the other is *not 0* then *X* could be a *0* and the output must be *X*. When one input is *1*, the other input 'dominates,' so *1* AND *R* gives *R* and *1* AND *F* gives *F*. If one input is rising and the other is falling, then there is no way of telling which will dominate. The result must be *X*, even though that may be a bit pessimistic. It can be argued that falling is tending towards *0* and so will dominate; but, if *R* had been present a 'long time' and *F* and just begun, this would be optimistic. *X* is safer. The OR gate is determined by similar arguments.

The second problem is what happens if a five-valued signal (say) appears on the input of an element whose model can only handle three values. In this case, the three-valued model requires a preprocessor to convert the *F*

Table 7.2. *Truth table for five-value AND and OR gates*

	AND					OR				
	0	1	X	R	F	0	1	X	R	F
0	0	0	0	0	0	0	1	X	R	F
1	0	1	X	R	F	1	1	1	1	1
X	0	X	X	X	X	X	1	X	X	X
R	0	R	X	R	X	R	1	X	R	X
F	0	F	X	X	F	F	1	X	X	F

and R values to Xs. This is rather trivial. It is more difficult when converting a 46-valued signal for use by a three-valued model (Section 7.8). Even more difficult is converting a multiple-valued signal to a two-valued signal. For example, does an X become a 0 or a 1?

The opposite case where there is a three-valued signal on the input of a gate whose model handles 46 values is usually easier since the smaller set is usually a sub-set of the larger.

Languages such as VHDL allow other user selected logic 'types' such as integers, bit_vectors, etc. In Prog. 7.1 a TYPE state_1 was used. This was defined to have the possible values D_0, D_1 and P_1. These may also require 'buffers' to deal with differential timing and inertial delays. However, most uses of non-bit style signals are for high level design when timing is relatively crude. The necessary functions must still be present in the environment. If a type is truly user selected it may be necessary for the user to write the buffer model.

7.8 Signal strengths

The number of values that a signal may take can be further increased. In particular a value 'Z' is required to record when an output is high impedance. This is used either with devices designed for bus driving or for wire-OR/AND gates. For example, TTL open collector gates have no active pull-up mechanism.

Consider Fig. 7.20(a), which shows a TTL open collector gate. With both inputs at HIGH (H) the output is at LOW (L). When one input goes L the output is intended to go H, but there is nothing to supply charge to the stray capacitances to pull the wire H. In this instance the output remains L. The situation can be expressed by providing a signal value of **high impedance**, Z. Correct circuit operation requires the addition of a resistor as shown in Fig. 7.20(b). Now the output can become H.

The value Z is not strictly a logical value. It is more properly described as a **strength**. Consider the open collector gate with a resistive pull-up as shown in Fig. 7.20(b). The time that the output signal reaches the switching threshold of the load devices depends on the time constant, RC, of the circuit as indicated in Fig. 7.21. This is a parameter which cannot be

Fig. 7.20. Open collector gates.

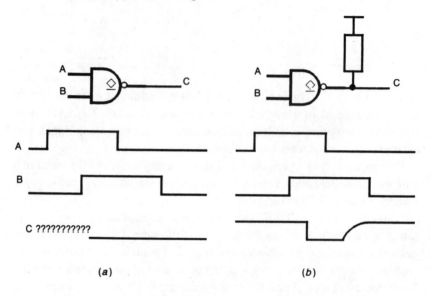

(a)   (b)

Fig. 7.21. Real waveform at open collector output.

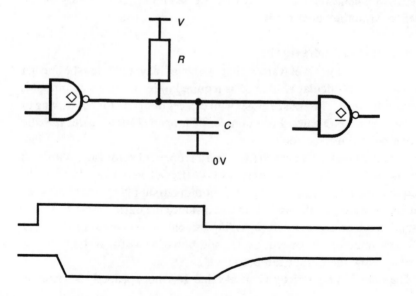

included in the basic gate model, since the value of C is design dependent and the value of R is user selectable. To handle this situation, the basic model of the gate can be specified with an external parameter. There can be several drive strengths chosen by the designer. These could be (for example) weak, strong, high impedance. The larger the number of strengths, the more complex is the simulation. In the case of the open collector gate, however, it is a question of selecting which of several delays to include in the run time model. That is a compile time action and need be done only once, but each instance of the open collector gate may then have a different model. The added delay must be held in the network memory and not in the evaluation routine.

VHDL does not specify any particular set of states and strengths. However, most implementations provide a variety of options within their standard PACKAGES (a PACKAGE is a means to specify types, functions etc. to be used within a design). In one common PACKAGE, three strengths are specified for use with a nine-value system. These are **forcing**, f, **resistive**, r and high impedance, Z. The nine values are

$$f0, f1, fX \qquad r0, r1, rX \qquad Z0, Z1, ZX$$

The term 'forcing' implies a low impedance driver or a power supply. $f1$ implies a strong 1 signal as opposed to a resistive 1 signal which will have a slower analogue rise time and hence a slower rise delay. Very recently (1992) a standard nine-value system has been approved by the IEEE.

Another VHDL PACKAGE uses a 12-value system which has an 'unknown' strength added to the nine-value system. Yet another PACKAGE has a 46-value system. The latter makes use of **intervals**. A new strength, **weak** (W) is introduced between r and Z, and a value D which represents a node which cannot store charge or a disconnected network. Table 7.3 shows the 46 intervals. There are nine values across the top of the table representing all possibilities from $f0$ to $f1$. The 46 values are shown on the left hand side with the range indicated in the table. Thus, for example, $fZ0$ means any state between forcing_0 and high impedance_0. That is, the state is known to be 0, but the strength is not known. It is one of the available strengths. Notice that $fZ0$ does *not* mean 'forcing high impedance 0,' which is a contradiction in terms.

Use of signal strengths is of primary importance in the simulation of MOS circuits. For simple MOS circuits, the speed of the gate depends on the sizing of the transistors. A short, fat transistor can conduct more current than a long, thin one, and so will charge capacitances more rapidly. Further, in an NMOS inverter, for example, the pull-up transistor is 'on' all the time. What determines which transistor 'wins' when the pull-down

Table 7.3. *46 strengths of VHDL*

	f0	r0	W0	Z0	D	Z1	W1	r1	f1
U	←	—	—	—	—	—	—	—	→
D					←	→			
Z0				←	→				
Z1						←	→		
ZDX				←	—	→			
DZX					←	→			
ZX				←	—	→			
W0			←	→					
W1						←	→		
WZ0			←	—	—	→			
WZ1						←	→		
WDX			←	—	→				
DWX					←	—	→		
WZX			←	—	—	—	→		
ZWX				←	—	—	→		
WX			←	—	—	—	→		
r0		←	→						
r1							←	→	
rW0		←	—	→					
rW1						←	—	→	
rZ0		←	—	—	→				
rZ1						←	—	→	
rDX		←	—	—	→				
DrX					←	—	—	→	
rZX		←	—	—	—	→			
ZrX				←	—	—	—	→	
rWX		←	—	—	—	—	→		
WrX			←	—	—	—	—	→	
rX		←	—	—	—	—	—	→	
f0	←	→							
f1								←	→
fr0	←	—	→						
fr1							←	—	→
fW0	←	—	—	→					
fW1						←	—	—	→
fZ0	←	—	—	—	→				
fZ1						←	—	—	→
fDX	←	—	—	—	—	→			
DfX					←	—	—	—	→
fZX	←	—	—	—	—	→			
ZfX				←	—	—	—	—	→
fWX	←	—	—	—	—	—	→		
WfX			←	—	—	—	—	—	→
frX	←	—	—	—	—	—	—	→	
rfX		←	—	—	—	—	—	—	→
fX	←	—	—	—	—	—	—	—	→

transistor comes on is which transistor can conduct more current. By designing the pull-down to be short and fat, relatively speaking, it will 'win' as required. Put another way, the pull-down transistor has a greater strength.

In switch level simulation the system is simulated at the transistor level, but the transistors are modelled as switches which are conducting or not. The solution of partial differential equations is not necessary. One proposal, due to Bryant (1984), is to use 16 strengths ranging from infinitely strong (a power supply or primary input) to high impedance (Z).

Other devices have both active pull-up and pull-down drivers but can have both switched off. These are usually used for multiple sourced buses where exactly one driver should be 'on' at any one time. Again, if all drivers go 'off,' then the output value is undefined and is recorded by the simulator as Z.

7.9   Towards a model for a flip-flop

7.9.1   More complex models

Complex elements, particularly ones invented by the designer, will have to be modelled by him/her as well. These could be higher level models of an ASIC for use in system design, or models for macros for use in chip design. Whilst problems of unequal rising and falling delays and inertial delays can be left to the environment, there are several other features which cannot. For example, in a flip-flop the preset and clear inputs override the clock. Some means for switching between two different evaluations is needed. The flip-flop will be used as an example of a more complex logical element. It cannot, in general, be modelled as a set of (usually six) gates, since that does not properly reflect the timing as opposed to the logical function. It will be seen that the state machine technique proves very useful.

No claims are made for the completeness of this model. In particular, some users might prefer to separate the error states into more different types than those suggested here. Readers not wishing to cover the full detail may omit all but a paragraph or two of Section 7.9.3. The rest should be scanned at least.

7.9.2   The 74xx74 style flip-flop

The models considered so far are either single input, or multiple input where all inputs are equivalent so that the buffer can be preceded by a simple gate. The flip-flop has several inputs which act to some extent independently and with differing effects. There are also two outputs. The relative timing of the various inputs can lead to several error conditions. The VHDL of a 'simple' D-flip-flop with preset and clear written to test the

Table 7.4. *Timing summary for 74xx74 type of flip-flop*

	Q	NQ	
Preset active, clear inactive	*1*	*0*	
Clear active, preset inactive	*0*	*1*	
Preset and clear active	*X*	*X*	(could be *1 1*)
Preset and clear inactive AND			
D = *1*, active clock edge	*1*	*0*	
D = *0*, active clock edge	*0*	*1*	
Q = *1*: (Preset time $< t_{pmin}$)	*1*	*0*	
Q = *0*: Preset time $< t_{pmin}$	*X*	*X*	error flag
Q = *0*: (Clear time $< t_{cmin}$)	*0*	*1*	
Q = *1*: Clear time $< t_{cmin}$	*X*	*X*	error flag
Clock post-active time $< t_{Hmin}$	*X*	*X*	error flag
clock pre-active time $< t_{Lmin}$	*X*	*X*	error flag
(clock period $< t_{clpmin}$)	*X*	*X*	error flag
Set-up time violation[a]	*X*	*X*	error flag
Hold time violation[b]	*X*	*X*	error flag
Preset or clear recovery time violation[c]	*X*	*X*	error flag

[a] D must be constant $> t_{su}$ before active clock edge.
[b] D must be constant $> t_h$ after active clock edge.
[c] Preset and clear must be inactive $> t_r$ before an active clock edge.

model below extended to over 300 lines of code and is anything but simple. Even then, certain simplifications were included in relation to outputs becoming unknown. Fortunately no input change requires execution of more than a fraction of the code. However, it is clear that the model should be written in such a way as to minimise the amount of code executed. It must also be borne in mind that procedure and function calls can be very costly in execution time. It may be sensible to increase the size of code to minimise such calls. Knowledge of how the software is run is clearly of value. The following represents only one of several possible approaches to a three-value model of a D-flip-flop.

In this description a flip-flop similar to the 74xx74 is assumed (Section 1.5). Preset and clear are asynchronous inputs and override the effect of clock changes. Table 7.4 shows the effects and possible errors. The preset and clear recovery times are usually significant but rarely specified in the data books. All three of the clock times are also necessary, though the following model omits the 'period' one. In Table 7.4 the case of the preset and clear both being active is shown as leading to both outputs being *X*. In some cases this might be *1 1*, or might lead to oscillation. It depends on the precise circuit used.

The state diagram consists of two parts. One of these, the P–C mode, is operative when preset or clear or both are active, Fig. 7.22. The state moves from this section to what might be termed the clock controlled section when both preset and clear are inactive, Fig. 7.24. However, checks on the time between preset or clear becoming inactive and the active clock edge are required. Fig. 7.25 shows checks on the clock timing in the clock controlled mode.

7.9.3 The preset/clear (P_C) mode

Fig. 7.22 shows the P_C mode section. There are five states. The reset state, *D_LH*, implies Q = LOW (*0*) and Q̄ = HIGH (*1*). In a similar

Fig. 7.22. D-flip-flop P_C mode state diagram. Note: P = A means preset is active; P̄ means preset inactive.

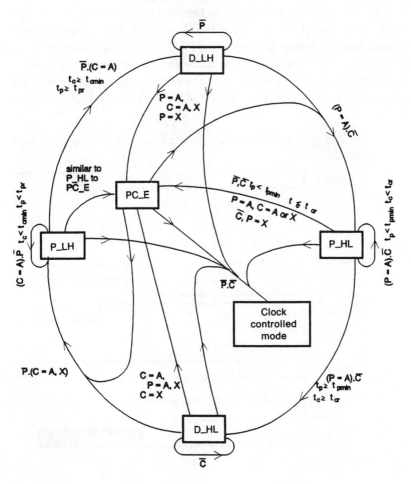

manner the other states are designated the set state (*D_HL*), possible_reset and possible_set states (*P_LH*, *P_HL*) an error state, preset/clear error (*PC_E*). If more detailed error reporting is required, the *PC_E* state could be split into several different states.

A flip-flop is normally in the clock controlled mode and will move to the P_C mode whenever preset or clear becomes active. Consider a flip-flop in the *D_LH* state (reset). The state will not change so long as preset is inactive ($\bar{P}$) (clock action is not being considered in this section). Suppose, now, that preset becomes active, Fig. 7.23. The mode changes to the P_C mode and the state to the *P_HL* state. The pseudo-variable *s* is planted in the event memory at time t_{pmin} ahead. This is the minimum width of preset pulse to guarantee an output change. T is recorded in the event memory associated with the preset data.

When *s* returns to re-activate the model, the state and time are checked. If the state is *P_HL* and T is t_{pmin} ago then the state becomes *D_HL* and output predictions of *H* and *L* for Q and $\bar{Q}$ are made, Fig. 7.23(*a*). If preset becomes inactive before this time (Fig. 7.23(*b*)) then the state becomes *PC_E*, the mode returns to clock controlled mode and predictions of outputs becoming *X* are made. When *s* returns the mode is not clock controlled so no action is taken. In the case shown in Fig. 7.23(*a*), when *s*

Fig. 7.23. Operation in P_C mode – Preset: (*a*) long and (*b*) short preset pulse.

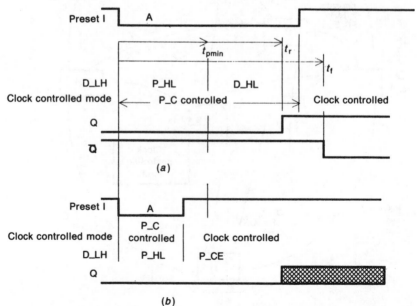

returns it is necessary to check that clear has been inactive long enough, as well as preset active. This covers the circumstance where preset and clear were both active and clear becoming inactive has changed the state from *PC_E* to *P_HL*.

The description in the previous paragraph presumes that the minimum width preset pulse is less than the output delay, which is usually the case. If it is not then the output will change to *HH* and possibly then *HL* ($t_r < t_f$), but the state will remain *P_HL*. If the pulse width error then occurs, the state changes to *PC_E* (preset or clear error) and the outputs are both changed to *X* after a further t_{fmin}.

If the flip-flop is in the *D_LH* state and preset becomes *X* then the state becomes *PC_E*, since it is not known whether preset is *0* or oscillating. The state will also become *PC_E* if preset is active and clear is either active or at *X*, since, in the pessimistic case, *X* can be *0*. However, clear being active or *X* while preset is inactive will have no effect on the *D_LH* state.

In running simulations, it might be required to stop when any error state is reached. However, it may be that there are several tests in one run and that a control reset is done between tests. In the case of data path flip-flops, 'unknown' data at some times may not matter. Hence the user will often want the simulation to continue after such 'errors.' It is for this reason that the errors are states in the state diagram rather than simply reports.

Fig. 7.22 shows how. If the flip-flop gets into the error state then preset and clear both inactive will still return the flip-flop to the clock controlled mode (see also Fig. 7.23). An active clock with a D input which is *0* or *1* places a 'good' value on the outputs. A second possibility is where PC̄ (or P̄C) returns the outputs to a known state.

If the flip-flop is in the P_C mode and *D_HL* state then a corresponding description applies, with clear replacing preset and the *P_LH* state replacing *P_HL*.

7.9.4   The clock controlled mode

Figs. 7.24 and 7.25 show a possible state diagram for the clock controlled mode. Fig. 7.24 is the main diagram and Fig. 7.25 adds some states for handling clock errors and an *X* on the D input. They presume that preset and clear remain inactive. From the previous description, it should be possible to follow the diagrams. It is presumed that both set-up and hold times are positive (see below). A second pseudo-variable is needed, since a pseudo-variable initiated in the P_C mode can return while in the C_C mode and must be distinguished from a pseudo-variable initiated while in the C_C mode. The clock error state, *E_CK*, is only relevant if there was a significant event elsewhere. For example, if the flip-flop was in the *D_LH* state, $D=0$ and there is an active clock edge, an attempt is being made to reset a flip-flop which is already reset. A clock width error does not matter in this circumstance. Once again it is possible to move from an error state to a normal state with appropriate input patterns. The CK_X state was found to be necessary to distinguish when a clock had gone to the X value and then to a non-X value. On the latter change, the state becomes E_CK and will revert to a non-error state with appropriate inputs.

Fig. 7.24. D-flip-flop clock controlled mode state diagram. t_{pr}/t_{cr} are preset/clear recovery times. E_AR is error-asynchronous recovery.

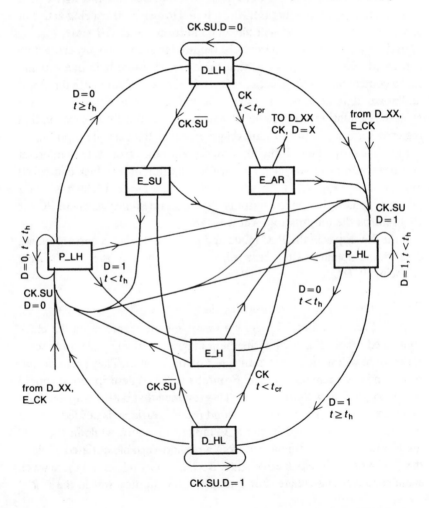

7.9.5 Timing errors

Checking of timing errors requires each input signal in the network memory to have an associated time value. For example, the D input changing has no effect on the operation of the flip-flop, but the time is recorded. Further, the time must be checked against the time of the last active change of the clock to ensure that hold times are obeyed. This will happen when in the *P_HL* or *P_LH* states, it being assumed here that hold time is less than the delay from active clock edge to output change. When the clock changes an 's' style prediction, v, is made for a time t_h ahead. When the model is re-activated and the time difference is at least the hold time then a change of state is confirmed. If the time difference is less than the hold time then a hold time error (*E_H*) state is entered and predictions of outputs changing to X are made (see Table 7.4). Similarly, when an active clock change takes place, the current time is first compared with the time of the last change of D as well as its value. If the time difference is less than the set-up time then the set-up time error state, *E_SU*, is entered and, again, outputs are predicted to change to X. Set-up time errors are regardless of the state of the flip-flop or value of D since the value of D has changed and the output cannot be guaranteed.

A number of error conditions are not considered here. These include

- an active clock edge while in the *P_HL* or *P_LH* state. That would require a very fast clock relative to the speed of the flip-flop, so is unlikely,
- multiple errors.

Fig. 7.25. D-flip-flop clock controlled error states.

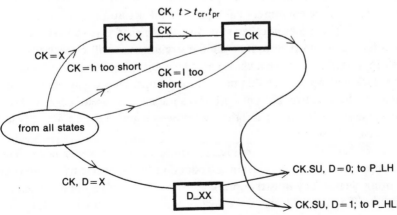

It is also assumed that if $D = X$ and an active clock edge occurs then the state changes to D_XX without checking for set-up or hold errors.

When an active clock edge occurs (preset and clear inactive) the current time must be compared with the time held in the network memory for the last change of preset and clear. If either of these differences is less than the P/C recovery time then the system enters the asynchronous reset error state, E_AR. The times of clock changes must also be compared with the time of the *two* previous clock changes to ensure that the *HIGH* and *LOW* times and the period are longer than the minima specified.

All errors must be reported to the user. These may be placed in the normal output stream together with the waveform information. This is dangerous, as the errors are then easily missed in a mass of data. It is better to have a separate output stream. In either case, an indication of the type of error is required. Thus the user needs to know not just the time the error occurred, but the route by which the model reached the error state and the last time at which each of the offending waveforms changed. For example, a short preset pulse should report the time of both edges of the pulse, and a set-up time error should report the time of both the D change and active clock edge. This information is not easily available unless the designer of the simulator has been careful to ensure that the relevant time from the past can be read from the network memory as well as the value.

Fig. 7.25 presumes set-up and hold times are both positive. This will be true of a 'simple' flip-flop. However, in more complex synchronous devices, it may not be so. Consider the device illustrated in Fig. 7.26(*a*), which might be an arithmetic unit with a registered output. Presume for the present that the hold time of the register itself is zero. The *minimum* time through the combinational logic may be (say) 20 ns, so the data inputs can be removed 20 ns before the clock edge and the logic still works correctly. In this case the hold time is specified as − 20 ns. Clearly the set-up time is the *maximum* time through the combinational logic *plus* the set-up time of the register.

Fig. 7.26(*b*) shows another situation in which the set-up time of the flip-flops is assumed zero. By a similar argument, it will be seen that the set-up time of the arrangement is negative and the hold time is the delay of the clock buffer plus the hold time of the flip-flops. Thus set-up time or hold time can be negative, but not both. To simulate the case of negative hold time, the model is written as if there were two set-up times and similarly for negative set-up times.

More complex devices obviously require more complex models. The examples given here provide an introduction to the problems involved in writing them. They illustrate problems for combinational and sequential circuits. The latter illustrate the sort of errors which may be found.

These sections have also introduced the idea of an 'unknown' value, possibly in several 'flavours.' Different CAD suppliers use different numbers of values and for different purposes. It is up to the user to take advantage of what is provided.

7.10 High level modelling

7.10.1 Behavioural models

So far the discussion has covered models of simple gates and flip-flops. 'Simple' gates may or may not include the not equivalence gate, depending on the underlying circuitry. The question now arises as to whether the whole system should be described in terms of AND, OR, NAND, NOR and buffers in order to get full accuracy. Such a process is known as **flattening** to gate level, since in the simulation only models of these simple gates are used. The network model is *structural*.

The second question to be raised is how to create models of complex devices. Even a 4-bit ALU, shift register or counter requires up to 100 or more simple gates. In practice, the underlying construction may be current steering logic, which makes 'gate equivalent' logic diagrams inaccurate from a timing point of view. It should be understood that this does not apply only to ECL logic. Some complex TTL logic uses current steering techniques internally (e.g. 7474 flip-flop!), and use of MOS pass transistors has a corresponding effect.

Many early simulators (pre-1990) could not handle anything other than simple gates and flip-flops, though most claim to handle not equivalent

Fig. 7.26. Negative set-up (*a*) and hold (*b*) times.

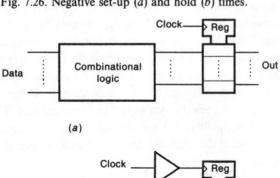

(*a*)

(*b*)

gates. Models used for complex devices flattened them to simple devices. This fills up the memory space available for storing the network and increases the number of events that can occur, thus slowing down the simulation. For example, a 4-but ALU has around 60 gates in its logic diagram. These will require network memory storage for 60 + elements with an average of three or four inputs and one output. A single input change may lead to between five and 10 events, each requiring a call of a model and, say, 10 instructions to execute, excluding those needed to manipulate the data structures.

Suppose, instead, that a behavioural model of the ALU is written. It requires network memory storage for only one element, though with seven outputs and 15 inputs. One might reasonably expect a reduction in storage space of between 10 and 20 times. Furthermore the change of any input *or any number of inputs* results in only one call of the model. It may require 50 instructions to evaluate the model, and rather more data structure manipulation than for *one* simple gate. Allowing for the cost of procedure calls, the high level model should run much faster than the gate level one. This advantage increases when several inputs change together, as may frequently happen in the simulation.

This leads to two things. Firstly, high level models run faster and require less storage. Hence larger systems can be simulated with given limited resources. Furthermore, there is no point in simulating at gate level when it is required to check the interactions of major blocks (Harding 1989). Harding also points out that many ASIC designers never check their devices in a system, and frequently they do not work in the system.

Secondly, it may well be necessary to simulate any logical element at gate level or even lower levels. If the logical element can have two or more models it is possible to simulate parts of a design at a detailed level and parts at a less detailed level. This again makes it possible to simulate larger systems with given resources. It also makes it possible to simulate parts of a system at different stages of design yet still within the system. For example, if a detailed design of a shift register has been completed but only the specification of the ALU, then the system can still be simulated. This is the intention of the VHDL facility, to write several architectures for the same piece of logic and to select the relevant one for a particular simulation using a configuration.

7.10.2 Hierarchical models – structural

The question now arises as to how to design a gate level model of a complex device. It is clearly very wasteful of people's time and energy to design every model from scratch. Take, for example, a shift register. This is made up of gates and flip-flops. If the gate and flip-flop models have been written in such a way as to enable input numbers and types, delays etc. to be passed to the model at compile time as parameters from the network, then it is only necessary to have these basic models. Similarly, if a single shift register stage is available, it can be used to build up longer shift registers. The problem now is how to specify the 'internal' connections of the longer shift register, and how to call the lower level models.

Consider Fig. 7.27, which might be a 32-bit ALU. On the left hand side is a list of inputs and outputs of the top level model as received from the network memory. This list will contain other information such as pseudo-variables, times associated with signals etc. These can be included without altering what follows in principle. They are ignored only to simplify the description.

The model of the 32-bit ALU will call a 4-bit ALU eight times. The 32-bit model 'knows' which of its inputs and outputs belong to a particular call of the 4-bit model and passes the appropriate data. The 4-bit ALU may then wish to call an AND gate model. Again, which inputs to pass are 'known' to the 4-bit ALU model. The AND gate will generate an output prediction. This *must* be a signal recognised by the *32-bit* model if it is to be sent to the event memory, since anything outside the model evaluator has to be related to that. Thus, if this gate is 'internal' to the 32-bit ALU, it must be represented in the data on the left hand side and possibly the centre column as a pseudo-variable.

The difference between a behavioural model and a hierarchical structural

Fig. 7.27. Hierarchical model.

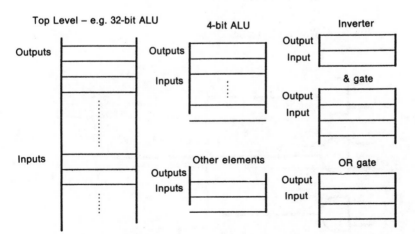

model must be emphasised. The behavioural model has the operation described in abstract terms as the manipulation of the inputs and possibly pseudo-variables to produce the outputs. The structural model uses low level elements and is used as a means to simplify the detailed description of a complex element. It is comparable to the use of library procedures in programming.

7.11   Wire gates

Open collector TTL gates, ECL gates (open emitter) and tristate devices are all intended for use with 'wired logic'. By this is implied a logic function constructed by wiring two or more output signals together. Fig. 7.28 is an example. If the logic family is TTL, then the logic function of the wire is an AND, since in this technology any real gate output pulling to *0* will force the wire to *0*. The AND pseudo-gate is shown dotted.

In theory, a wire gate has zero delay and so cannot be used in the usual way with the affected components list. Furthermore, multiple input changes at the same time are likely. The situation may be helped somewhat when the connections are metal, since the rising and falling delays will be the same and problems of cancelling predictions are not present.

The wire gate has a logic function. Thus the network compiler must do several things.

- Recognise that several gate outputs are connected together.
- Check that these outputs are compatible and the connections are legal: for example, conventional totem pole outputs cannot be wired together.
- Insert into the network description a pseudo-gate with appropriate function and values.

Fig. 7.28. Wired AND gate.

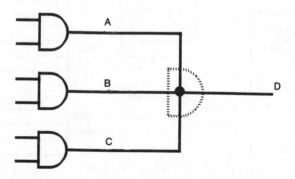

- Insert appropriate entries into the fan-out table which represents the interconnection of the gates.

  (a) The fan-out memory indicates just one load for each of A, B and C in Fig. 7.29.

  (b) The fan-out memory indicates one output and a set of driven inputs for D.

Suppose that there is a change on an input to gate A in Fig. 7.29. The output prediction for A eventually appears from the event memory. The fan-out memory passes this to the pseudo-gate, D, in the affected components list and it is eventually evaluated.

Evaluation of gate D must take into account the values on B and C. If A is *1* and B is *0*, there is an incompatibility and an output X and an error must result. The only legal situation is for one and only one of the inputs to be at a logical value. The others must be Z or resistive. This implies that the signals A, B and C as shown in Fig. 7.29 cannot be the same which, of course, they are. However, when the system reports values for them, the value D should be reported as well.

Prediction of D is made. Although the gate has zero delay in theory, in practice connecting several outputs together adds capacitive load, and to give this gate a unit delay is not unreasonable. In VHDL it should make use of deltas if the delay is genuinely small.

The use of wire gates must be done very carefully.

- The wired outputs must be physically close together – the longest distance apart should be less than the distance travelled by a signal in half the signal rise time. This must be checked by a rule checking program after layout. If the rule is not observed, reflections will occur, resulting in loss of noise immunity or even oscillation.
- If the gate is actually made using MOS pass transistors then the situation is more complex. Multiple state switch level simulation will be necessary. Further discussion is beyond the scope of this book.

VHDL has provision for the writing of resolution functions for wired gates. The above represents an approach to their implementation in the simulator (environment).

7.12 Hard models

7.12.1 Non-memory devices

Many systems to be simulated will contain very complex devices such as microprocessors, large memories etc. It has been suggested that to write a model of such a device would take as long as to design the device itself. This is likely, as the two processes are similar. Furthermore, the specification of a complex device does not record all the behaviour under unusual or exceptional conditions which can occur in simulation.

To alleviate such problems, a number of companies have produced hardware modellers. These are special pieces of hardware which can be connected to a conventional computer. The cards contain sockets which can hold one or more physical devices which perform as the 'model' in the simulation. The simulator is written in such a way that, when designated devices are affected in the affected components list, the relevant signals on the computer interface are driven. The hardware modeller samples the device output signals at regular intervals and sends the result back to the computer where the interface software will place the appropriate predictions in the event memory.

Hardware models are functional models. They cannot contain the necessary timing parameters or checks since the actual device used can never be 'typical' or 'worst case' or anything else 'special.' Any timing checks must still be performed in a software model. This may be run on the modeller or in the normal simulator. Nevertheless, with sufficiently fine sampling of the outputs, reactions to stimuli other than the static values can be obtained.

A major problem with a hardware modeller, as with any hardware, is its lack of flexibility. Building hardware to be compatible with many host computers and with different simulators is not easy. Two approaches have been used.

Logic Modelling have built a machine, the LM1000, which communicates via a standard, high speed link, Ethernet, with the host computer. Software has been produced to enable the modeller to work with several operating systems including flavours of UNIX and DOS. The software in the interface makes as little use of operating system facilities as possible in order to ensure maximum portability. Only a sub-set of the usual Ethernet TCP/IP calls are needed. This has the added bonus of extra speed in processing.

The second approach is exemplified by the Dazix Physical Modelling Extension, PMX. This runs only with the Dazix (Daisy) hardware and software which makes it more difficult to use with simulators from other manufacturers. Again, the PMX uses Ethernet connections.

The two systems are similar in construction, the LM1000 (Kelly *et al.* 1989) being a little more flexible. It can handle up to 32 devices, each with up to 80 pins in an 8 by 4 matrix. Groups of two, three or four devices in a 'column' of sockets can be coupled together to allow devices with up to 320 pins to be simulated (Fig. 7.29). Patterns can be presented at up to 25 MHz with ± 1 ns or $\pm 1\%$ accuracy. Normally output samples are taken based on the clock and delays specified in the 'core' model code. In one mode, the timing measurement strobe can be swept to look for input and output changes with resolution down to 0.5 ns.

Software within the modeller itself performs a number of functions. These include configuring the hardware to know which socket pins are inputs, outputs or bidirectional pins on the device. The modeller also performs timing checks on the devices, but, as indicated above, this is software within the modeller CPU.

A device of only combinational logic is easily simulated on a hardware modeller by applying a pattern to the inputs and observing the outputs an appropriate time later. For a device containing registers, it will be necessary to apply several clocks to reach a particular place in the simulation. In the worst case, with certain devices based on dynamic logic, it will be necessary to apply all the input vectors that have gone previously every time the device model is called. If this is not done, then there is a chance that the device will not be in the required state prior to application of the 'new' vector. Fortunately only a very few devices come into this category.

- Memories should not use such a modeller (see below).
- Most CPUs require the clock to run continuously. If the modeller presents a NO-OP when a new pattern is needed, the long set of old patterns is not required.

Fig. 7.29. LM1000 architecture; 4 by 4 devices.

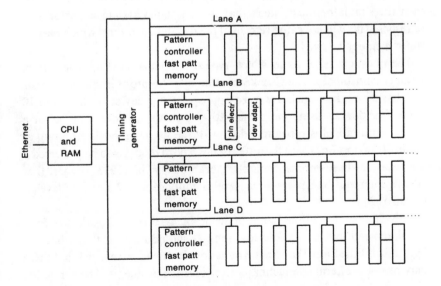

However, there is another potential problem. Suppose a system includes two (or more) identical devices to be modelled on the hardware modeller – a CPU, for example. Suppose that for some reason – size of the modeller, number of other devices using the modeller – it is not possible to have two CPUs on the modeller. Then, when it is required to simulate CPU_2 having presented a pattern to CPU_1, the whole history of CPU_2 inputs must be presented to ensure the correct initial state. The same procedure must happen every time it is required to change from one instance of the CPU to the other. With long simulations this can be slow. To avoid having to send all the data from the host every time, the modeller contains a **pattern memory** for each pin. The LM1000 supports sufficient memory for 256K patterns.

7.12.2 Memory

Memory could be simulated on a hardware modeller. However, to simulate a sensible system, a rather large number of chips are needed. It is best to arrange to do the functional simulation of memory with a bank of memory in the main store of the host computer. This memory holds the data. A network memory entry will hold other information, such as times of last change of each signal and a pointer to the start of the data memory. The model evaluation routines will perform the necessary timing checks. As the data is held in real memory, it is possible for simulated static memory to have its data held in dynamic real memory or vice versa. In the former case the refresh cycles for the real memory are performed by the normal memory controller, not by the simulator. If the dynamic memory is being simulated then the simulator must check that the simulated system is performing refreshes. This will *not* cause refresh of the real data memory, which may be static memory anyway.

The question arises as to which type of model is most appropriate in given circumstances. The QuickSim manual of Mentor Graphics suggests that basic built in models are most appropriate for devices of less than 10 gates. For devices using more than 10 gates, either a library model or a behavioural model is recommended, with the library models stopping at about 200 gates. For more than 1000 gates a hard model is to be preferred if the necessary hardware is available. Abramovici *et al.* (1983) suggest that 10 min computation at functional level is equivalent to 2 h at gate level, a factor of 24.

This chapter has outlined some of the problems of modelling devices for a simulator and indicated some solutions. It should be clear that what has been said is little more than an introduction. It should be realised that there may be many alternative techniques to solve the problems. The simulator designer has to assess the value of each technique to find the 'best' for the particular implementation circumstances. The problems are so great that there are now whole companies devoted solely to writing models for simulators.

8

Timing verification

8.1   Introduction

Although the event driven simulator allows timing to be included in a simulation, it is extremely difficult to devise a set of tests that would show up all possible timing problems. Such a set of tests would have to analyse the network structure to find where two paths from the same signal converge later in the network. One of these would have to be assigned maximum delay and the other minimum. Such a situation was shown in Fig. 1.1 given that the two inputs were related and is known as reconvergent fan-out. The four-gate not equivalence example has five cases of reconvergent fan-out. A procedure is needed to find unwanted short pulses. It requires that all associated signals have the relevant states, which is why it can be difficult to drive. Having found a potential short pulse, it must be decided whether it matters. At the input of another gate, it does not. At the asynchronous input to a flip-flop, it most certainly does.

The second problem with timing is to be sure that the longest path, often known as the **critical path**, through a combinational network has been activated in order to ensure that the logic can operate within the design time specified. In particular, with synchronous logic, it is necessary to check that the logic works within the specified clock periods. The naive analysis of the four-gate not equivalence circuit designed earlier indicated the dangers of pattern sensitivity. That analysis was by no means complete (Section 6.4, last paragraph).

To ensure that adequate test patterns are generated is not impossible, but running the set of tests is very time consuming. Generating the set of tests can be done by examination of the structure or **topology** of the network. Having done this, it is possible to analyse the timing problems without having to generate tests for an event driven simulator or running such simulations. Such an analysis is known as **timing verification**. Timing verification is an attempt to simulate a network with all possible input patterns in one pass. *It begins with the assumption that the logical design is correct.* It must have been thoroughly simulated for functionality, therefore. There is no need to have included timing in that simulation, so a compiled code simulator could be used for speed. The timing verifier is used to check timing relationships only. It is faster than normal simulation because it is independent of the input patterns (but see later).

8.2   Computing the critical path

One way to tackle the problem of finding the longest path is to start from a primary output of the network and work backwards, assigning a maximum time to each signal as the analysis proceeds. Consider Fig. 8.1, another form of not equivalence logic. G is the primary output.

Suppose for simplicity that each gate has a delay of 10 ns. Signals E and F are both assigned a time of 10 ns delay. Signals C, D, A and B are then assigned a time of $10 + 10 = 20$ ns. Finally, tracing back from C and D, one gets to A and B again with a time of 30 ns. As 30 is greater than 20, the time associated with A and B is re-assigned to be 30 ns. The primary inputs have now been reached and there are no new paths to trace, so the longest path has been found.

This example is over simple for several reasons.

- In general the network will have several outputs and the number of possible paths back to the inputs will be very large. Furthermore, the **cones of influence** back from different outputs may overlap so that a signal on the longest path for one output is not on the longest path for another. The analysis can become very large, but not as difficult as solving the problem with an event driven simulator. Means to simplify the timing verifier problems have been researched.

Fig. 8.1. Circuit to illustrate timing analysis.

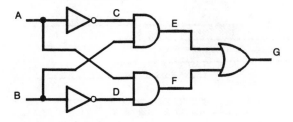

- Some paths are not possible and if included will lead to pessimistic results. These are called **false paths**, and are discussed in Section 8.5.
- There may be a need to distinguish between rising and falling delays (note that all waveforms still have infinitely fast rise and fall times). For example, if the rising and falling delays are 8 and 15 ns respectively and a signal passes through two inverters in series then the total delay is 23 ns for both directions of change, and not 16 ns or 30 ns.
- Fig. 8.2 illustrates another problem (HITIME). The flip-flops represent part of a shift register. Q_1 is *1* and all other signals are *0*. The clock now becomes active – goes to *1*. After buffering, the clock to the flip-flops could appear anywhere in the shaded area which represents the production spread of the buffer delay. The earliest time that Q_1 or Q_2 could change is $t_{buff\_min} + t_{ff\_min}$ and the latest $t_{buff\_max} + t_{ff\_max}$. It appears from the diagram that Q_1 could change before the clock to Q_2 changed. However, the user realises that whilst there is some uncertainty about when the buffer output changes, it changes at the same time for both flip-flops and the operation must be correct. This is an example of **common ambiguity**. A good timing verifier should not raise objections to this circuit.

From the above it will be realised that timing verification is not a panacea for all ills, and can be pessimistic. At this time (1992) it is likely that, even

Fig. 8.2. Illustration of common ambiguity.

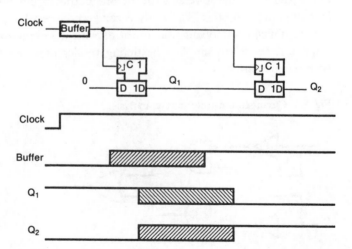

with means to eliminate most false paths and common ambiguity, it will still be pessimistic. It will be seen later that, without due care, it could be optimistic. It is important that this should not happen since that would lead to unwarranted confidence in the design and problems with commissioning and production testing that could only be solved properly by redesign.

In view of all these problems, it may be asked if timing verification is worthwhile. Some estimate of timing must be made. In particular, the longest path must be determined. It has been said that manual selection of the critical path is based on 'designer hunches,' the excellence of which will be highly dependent on experience or even luck, especially as most people involved with detailed design are relatively inexperienced. It is reported that on the RISC-1 project at Berkeley and on the MIPS project (Jouppi 1987) these estimates were out by a factor of *four*. The problem of false paths is the cost of neglecting real values propagating through the networks which would be observed in conventional simulation.

8.3 Methods of timing verification

8.3.1 Path enumeration

The procedure for calculating delays illustrated in association with Fig. 8.1 is known as **path enumeration**. The 'start point' is either a primary output, or, in the case of synchronous logic, the output of a combinational logic block feeding a register. The terminal point will be either the primary inputs or the clock signal to a register feeding the relevant combinational logic block. The clock is the appropriate point since the delay of the flip-flops making up the register must be included. In the example, the circuit was effectively levelised and all delays at a particular level found before moving back. With blocks of logic with very different delays, this will be difficult to do precisely, but does not matter, since at each stage, if a delay figure larger than that already found for a node is calculated, the larger figure replaces the smaller as happened with A and B in the example.

An alternative approach would be to follow one path right back to the input. Considering again Fig. 8.1, this might be G–E–B. It is necessary at G and E to note that they have two inputs. On reaching a primary input the program returns to E and traces that input back to A. It then has to return to G to trace the paths F–D–B and then F–A. This is known as a **depth first** approach. By contrast, the previous approach is referred to as **breadth first**.

The path enumeration procedure suffers from the large number of paths which, in turn, leads to long run times. It has the advantage that the time for every node is known. It is also possible to tell the software to ignore certain paths, thus avoiding the problem of an obvious false path. However, that relies on the user recognising the false paths first, which is non-trivial, not to say error prone for a large network.

8.3.2   Block orientated path trace

In this method of finding paths, one begins at the start of a block – the primary inputs or a register clock. The delay to each output is then calculated in either breadth first or depth first mode. Longest and shortest paths can be calculated by taking the higher or lower value respectively on second or subsequent calculation at a particular node. In this case, all the outputs of the block are calculated at least once.

It is claimed that this approach is faster than path enumeration. However, although it can calculate the longest and shortest paths, the earlier versions, at least, lost the details of how the critical path was built up (Ousterhout 1985). As this is moving forward through the network to all outputs, telling the system to cut out a false path to one output may cut out paths to other outputs which are not false. Hence the length of the critical path tends to be very pessimistic.

8.4   Description of the network

The network is usually reduced to a directed graph (Perremans *et al.* 1989) in which the nodes are the signals in the network and the edges are delays. Fig. 8.3(*a*) shows a circuit and (*b*) an appropriate graph. The number in the gate is used to identify the gate, and also its delay, for the

Fig. 8.3. Timing verification example: (*a*) circuit and (*b*) directed graph.

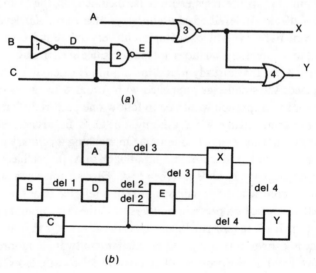

(a)

(b)

Table 8.1. *Data structure for Fig. 8.3*

Address	Signal	Comment	Values			
1	A	max delay	0			
2		back ptr	primary input A			
3		forward ptr		X	4	
4	X	max delay	3̶ 6			
5		back ptr		A̶ 1̶	E	17
6		forward ptr	primary output			
7		forward ptr		Y	8	
8	Y	max delay	7 10			
9		back ptr		X	4	
10		forward ptr	primary output			
11	B	max delay	0			
12		back ptr	primary input B			
13		forward ptr		D	14	
14	D	max delay	1			
15		back ptr		B	11	
16		forward ptr		E	17	
17	E	max delay	3			
18		back ptr		D	14	
19		forward ptr		X	4	
20	C	max delay	0			
21		back ptr	primary input C			
22		forward ptr		E	17	
23		forward ptr		Y	8	

purposes of the following example. Data associated with each node should include the maximum delay from an input to that node. It may contain the minimum as well. It should also contain details of the path to that node for the maximum delay, or at least sufficient information for that to be computed. A suitable back pointer might be sufficient.

Table 8.1 represents a possible data structure for Fig. 8.3. Starting with A, the maximum delay is zero as this is a primary input, and there is no back pointer. There is just one fan-out, so the forward pointer is to the next store location, 4. Moving forward in depth first mode, location 4 represents signal X. The delay is 0 from location 1 (A) plus 3 from the OR gate, namely, 3. For the moment ignore the fact that this is crossed out in the table. The delay came from the path from A, so the back pointer is to line 1. X is a primary output and is marked as such. It is also an input to Y so there is a second forward pointer which can be set to the next free location, namely, 8.

Proceeding to Y, the delay is $3 + 4 = 7$, the back pointer is to X (4) and the forward pointer is a primary output marker. There are no further fan-outs to follow so the procedure returns to the next primary input, B. The table continues to be built in a similar manner. However, the forward pointer from E in line 19 is to X, which is in line 4. The delay at E is 3 and this must be added to the delay of the OR gate forming X, giving a total of 6. Line 4 currently records a delay of 3. As 6 is geater than 3, the 3 is replaced. This, in turn, means that the back pointer from X for the longest path, which was previously to A, must be replaced by a pointer to E, 17 as shown in Table 8.1. The path forward to Y must also be recalculated so far as delay is concerned.

In constructing the structure for C to E it is found that the delay to E is $0+2=2$, which is less than the 3 recorded in line 17 for E. That path can stop, therefore. A similar argument applies to the path C to Y.

Clearly other structures than this are possible. Equally this structure can be expanded to include other information. Instead of forward and back pointers, it would be possible to contemplate a pointer into another piece of memory holding the complete longest (and shortest) path for each node. This might be built after a structure like this one, since maintaining it as it was being built might be quite difficult if a whole path has to be replaced rather than a single pointer as in the example.

8.5   False paths

Consider Fig. 8.4. A signal may pass through two paths of delay 10 and 20 units to a multiplexer. The output of the multiplexer then passes through two more paths to a second multiplexer. A simple analysis of this arrangement which ignores the functionality will find four possible paths with delays of 20, 30, 30 and 40 units. However, it will be observed from the connection of the control to the multiplexer that, *provided that* the control signal is fixed and a signal passes from the 10 unit delay to the first multiplexer, it cannot pass through the second 10 unit delay to the output of

Fig. 8.4. Illustration of false paths.

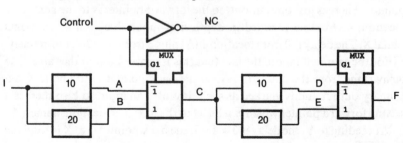

the second. Similarly, the 20 plus 20 unit path is not possible. The 20 and 40 unit delay paths are known as **false paths** since they cannot occur in practice.

A second example is a two-phase latched system as found in many MOS systems. Fig. 8.5 shows two sets of combinational logic separated by latches. Data cannot pass through many stages, since if ϕ_1 allows latch L_1 to be open then latch L_2 is closed and vice versa. The longest path is from the start of ϕ_1 through L_1 and logic_1 to the *end* of ϕ_2 *less* the L_2 set-up time.

There are two primary sources of false paths. Fig. 8.4 is an example of a **logic dependency** false path. These can usually be handled by 'case analysis.' Observing Fig. 8.4, the control is forced to *1* for one run and to *0* for a second. If the control signal is not easily 'controllable' and/or affects other paths, this could present serious problems. With many such controls, many timing verification runs could be necessary, and obtaining independence of one control from another may also be difficult. The mere fact of there being a difficulty could indicate a need for a redesign.

The other source of false paths is where there is uncertainty over the direction of signal flow, for example through a pass transistor network in a switch level simulation. This problem is usually solved by marking a direction of flow on each transistor record. This will be difficult to do automatically, and easy to get wrong by manual means.

It has been suggested that either false paths do not occur in a particular network or that they are numerous (McGeer and Brayton 1989). Arrange-

Fig. 8.5. A two phase latched system.

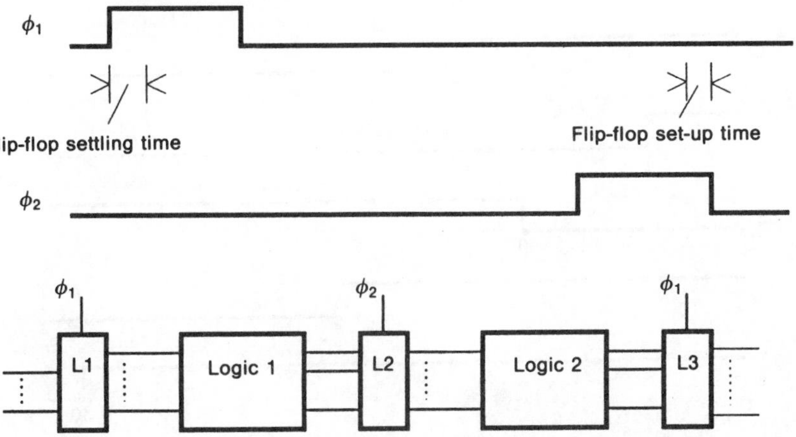

ments such as Fig. 8.4 will be common in some designs. In other cases, conditions where false paths could occur, mainly with reconvergent fan-out, should be avoided since there is a danger of producing brief pulses. A case in point is illustrated below.

False paths can be detected by making use of forward and backward traces in a manner similar to the D-algorithm. Consider Fig. 8.3(*a*) and the path B–D–E–X–Y. For the signal X to reach Y, signal C must be *0*. For signal E to reach X, A must be *0*. For D to reach E, C must be *1*. C has already been specified to be *0*, so the path B–D–E–X–Y is not sensitisable and could be labelled as a logical dependency false path. The longest delay will be assessed as 7 units (A–X–Y).

Consider, now, the waveforms shown in Fig. 8.6. B goes to *1* at time *0* and hence D goes to *0* at time 1. C is *1*, so this change can propagate to E at time 3. If C remained at *1*, the change would not reach Y since gate 4 is an OR and a *1* input will dominate. However, before the change starting at B can reach gate 4, C changes to *0*. That is not propagated to Y since X is still at *1*. Thus it is found that there is a delay of 10 units from B to Y, and the earlier analysis of this section was optimistic. Note that Table 8.1 gave the correct result.

This illustrates the difference between the static analysis represented by the basic D-algorithm and a dynamic analysis. It can be deduced that, if a 'control' signal such as C at gate 4 for path B–D–E–X–Y appears earlier than the signal on the path of interest then it should *not* be regarded as a

Fig. 8.6. Some waveforms for Fig. 8.3.

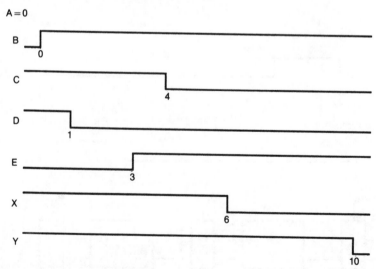

control signal. The reason is seen in Fig. 8.6, where control signal C at the input of E changed later than signal D coming from B, also at the input of E, but earlier than the signal at X which also originates at B. The longer path is opened.

The reader should analyse the path C–E–X–Y. The maximum delay is 9 ns.

A similar result is obtained by considering the network of Fig. 8.4 with the control changing. Suppose that the data input, I in Fig. 8.4, changes at time 0 and the control at time 25. Suppose that the device delays are all zero. Fig. 8.7 shows the waveforms. As the control is *1*, C follows B, at time 20. D goes to *1* at time 30 and E to *1* at time 40. F follows D until the control changes, after which it follows E. Hence F does not change until time 40. The data *appears* to take 40 ns to pass through the network. The false path is only false if the control input is steady. Thus this is a case of a **static** false path. Notice that the rule that if a control signal changes before a data path signal it should not be regarded as a control signal applies here, specifically the signal at the input to the second multiplexer in Fig. 8.4. The reader might like to examine what happens if real device delays are used, and the changes are not at 'convenient' times. In particular, the delay from a change on the control input of a multiplexer is usually different to the delay following a data input change. The output can become difficult to determine, and may well become unknown for some period.

The example of Fig. 8.3 and others that have appeared in the literature to illustrate this point are somewhat contrived. They so obviously have a

Fig. 8.7. Control change in Fig. 8.4.

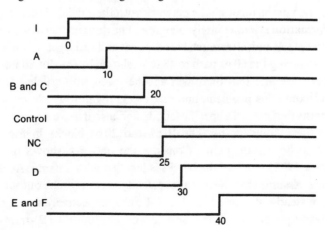

danger of spikes that no good designer would use them. In real circuits, such problems would occur accidentally over much larger sections of logic and would not be so easily spotted as problems. The example of Fig. 8.4 is quite likely to occur.

8.6 Use of timing verification

Look again at the purpose of the timing verifier in the light of the previous sections and see what data can be obtained from it.

- Is the longest path through the network less than that specified? This could be a clock period less set-up time for an edge triggered synchronous system, or ($\phi_1 + \phi_2$ − hold time) for a two-phase latched system.
- Is the shortest path through the logic block long enough to satisfy any (flip-flop) hold time criteria?
- Is it possible to generate short pulses in critical places such as the asynchronous inputs to flip-flops?
- Are two or more waveforms correctly positioned in time relative to one another to enable correct circuit operation. For example, are the RAS, CAS, read/write and various enable signals of a dynamic RAM in the right relative places? This includes latest time of one or several relative to the earliest times of the others.

If one determines that the critical (longest) path through the network is too long for the specification, it will be necessary to do some redesign. In this case, it is necessary to be able to extract the details of the path. However, this may not be sufficient. Suppose that one had a clock period of 50 ns (20 MHz) and that the critical path was found to be 60 ns. Fig. 8.8 shows two particular cases. Figures in the blocks represent delays and in each case two dashed lines show the two paths through the network with their delay. Only the critical path information is immediately available. The designer decides to work on the longest block and, after much blood, sweat and tears, gets the delay down to 18 ns, reducing the 60 ns path to 48 ns as shown by the dotted lines. Only then does (s)he find that there was another path with a delay of 55 ns.

To overcome this problem, one needs to know not just the critical path but *all* paths that exceed a specific length, and also if there are any common parts to these 'long' paths. The 10 ns and 20 ns blocks in Fig. 8.8 are common to both long paths. Consider the network shown in Fig. 8.9 (Hitchcock 1982). Each of the blocks is labelled with a name (A–H) and a delay time. Assume that the inputs are steady at time 0. The outputs of A, E and F are steady at times of 2, 3 and 4 units respectively and are shown above the output connection in the diagram. The inputs to B arrive at times

2 and 3, so the output will be steady at the later of $(2+4)$ and $(3+4)$, namely 7 units. The outputs of C and D are then at $(7+1)$ and $(8+3)$. The output of G is determined in a manner similar to B, namely max $((7+1),(4+1))=8$. The output of H is at time 13. Notice that all signals are now associated with a time but not with a path. It would be possible to create a linked list for each output, showing the longest path to that point. Thus B would have the

Fig. 8.8. Problem with shortening one path.

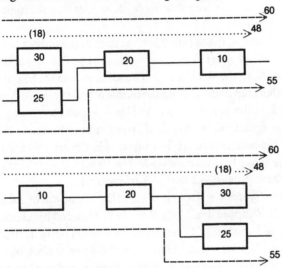

Fig. 8.9. Use of 'slack'.

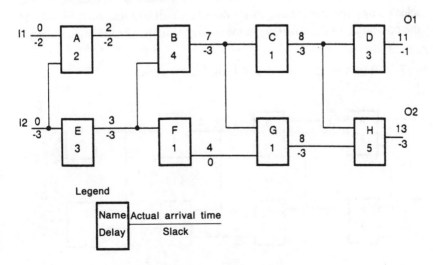

Legend

Name	Actual arrival time
Delay	Slack

list (E, I2) and H would have the list (C, B, E, I2) or (G, B, E, I2) or preferably both. Thus the path can be retained at a cost (see above).

Suppose, now, that the specification required both outputs to be present at time 10. Define a quantity called **slack**. If a signal arrived at the output of D at time 8 then it is 2 units earlier than necessary and there is 2 units of slack. Alternatively, the time of the signal at the output of D could be allowed to slip by 2 units and still be within specification. In fact, in the example, the output of D appears at 11 units, so the slack is $10-11=-1$ unit. This is shown below the connection line. Similarly the slack at the output of H is -3. A negative value for slack implies that the logic is slower than the specification.

Working backwards, the required arrival time at the output of C is either $10-3$ via D or $10-5$ via H. The *lower* of these is 5. The actual time at the output of C was 8, so the slack is -3. The same is true of the output of G.

Taking another step back, the required time at the output of B and F is 4. Hence the slack at the output of B is $4-7=-3$, and at the output of F is $4-4=0$. At the input to B, the required arrival time is $4-4=0$. This gives a slack of -2 at the output of A. At the output of E, the arrival time is either 0 for the route via B or 3 for the route via F. The lower is 0 and the slack is -3. The slack at the inputs is -2 and -3 respectively.

Observing the figures for slack, it is seen that the negative values funnel through block B. Suppose that block B could be replaced by a faster version having a delay of 1 unit (LSTTL replaced by AS, say). Fig. 8.10 shows the revised diagram. The output of D has a time of 8 and a slack of 2, and the output of H has a delay of 10 and slack of 0. Redesign of one block only has corrected both problems.

The difficulty with this approach is that the block to be redesigned was selected by pattern matching by the designer. This is much more difficult for the computer to do. Further, in much bigger networks the logic diagram

Fig. 8.10. Circuit of Fig. 8.9 after modification.

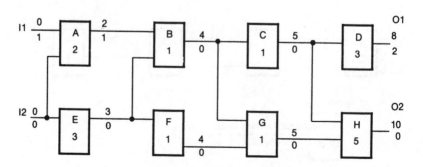

can rarely fit on a single sheet, and visual observation of the paths is virtually impossible. Nevertheless, the approach is interesting. The figures for slack do give valuable clues to what needs to be done. A suitable linked list print-out of several related paths may make the relevant paths visible.

There is another possible use for these techniques. Suppose that a design is produced. It could be by automatic synthesis from a higher level description. It is found by timing analysis that all the slacks are large. The designer (or the automatic software) can now ask if some or all the blocks of the design could be built from a slower technology, on the grounds that this type of circuit will use less power or be cheaper or both. This is particularly apposite in the design of large scale integrated circuits. Given a suitable map of the longest paths, appropriate blocks can be selected for modification, even if all blocks cannot be modified.

8.7   More complex models

It will be obvious that a trivial adaptation of the procedure just described will allow shortest paths to be found. Equally, if a minimum time through the network is found, then the slack will be defined to be (maximum arrival time calculated working forward) − (time specified for the output and traced back). With this definition, unacceptable slack is again negative.

It has already been indicated that, for logic units with different rise and fall delays, it is easy to get pessimistic results. It has also been suggested that some paths are not valid. To overcome these objections the SCALD (McWilliams 1980) verifier uses seven values. The values 0 and 1 could be used to force certain paths open or closed. That might mean running more than one pass of the verifier to obtain results under different conditions but the results would be less pessimistic. More seriously, with many such signals it will be as difficult to get a comprehensive set of tests as with the event driven simulator, since the timing verifier is now dependent on input values. Other values used by SCALD were stable, changing, rising, falling and unknown. Consider Fig. 8.11. The first number of a pair is the time of

Fig. 8.11. Use of rising and falling times.

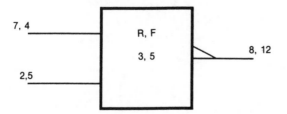

the rising transition and the second that of the falling transition (as for the models, all transitions are actually instantaneous). There are two inputs to the block which has an inverting output. Thus a falling input leads to a rising output and vice versa. The latest rising output occurs at max (falling input time + rising time), i.e. $\max((4+3),(5+3))=8$. The latest falling output is at time $\max((7+5),(2+5)=12$. Apart from the extra complexity here implied and the need to have some tables for evaluation of different gate types comparable to those of Chapter 7, the extension of the techniques is reasonably obvious.

9

Fault simulation

9.1   Introduction

Simulation has two purposes:

- to verify the design,
- to confirm that a test program will find faulty devices and systems.

In order to achieve the second of these aims, the failure modes of the system must be known. The effect of these failures must then be transferred to equivalent effects in the circuit models and tests generated to find these faults. It has already been said that a model based on wires stuck at 0 or s-a-1 has been found to be very useful and that an extension to include stuck-open faults of MOS circuits covers most other faults. This latter requires pairs of test vectors rather than single vectors. Finally, tests to find wires shorted together are needed. In practice, few test programs attempt to find these.

The number of possible faults is very large. For k nodes there are $2k$ single s-a faults and $3^k - 1$ faults including multiple faults. The number of bridging faults will be dependent on layout. Chapter 3 discussed design techniques that enable the number of tests needed to be reduced by partitioning large designs and test time to be reduced by permitting parallel testing of partitions. Even so, most test programs only aim to test for single s-a faults and many systems have test programs which cover only 80–90% of these.

Is this justifiable? Williams and Brown (1981) derived an empirical rule which said that, for a process yield of Y and fault coverage of T, the number of faulty chips accepted as good (defect level) is

$$\text{defect level} = 1 - Y^{1-T}$$

For a fault coverage of 99% and a manufacturing process with a 50% yield, 1% of faulty chips are accepted as good. If the yield is 10% then over 2% of faulty chips are missed, and if the yield is 50% and the fault coverage is 90%

then 7% of faulty chips are accepted as good. If testing is for single faults only then these figures will be worse – 3 to 4% has been suggested for the first case[1] (Hodge 1990). In the specific case of IC testing, whole areas of a wafer are bad (or good) and hence within a 'bad' area, multiple faults on a chip are likely. At a system level, it is possible that many on-chip multiple faults will appear as single faults at the terminals. The conclusion from this is that the lower the yield of the manufacturing process, the higher is the required fault coverage. Usually a test program that finds 80% of single s-a faults is simply unacceptable and some attempt to find multiple faults is highly desirable.

In spite of this the majority of authors limit themselves to single faults. Single fault cover may well relate to multiple fault cover. For example, Agarwal and Fung (1981) state that for fan-out free circuits (see below) 98% of multiple faults up to six at a time will be detected if single fault cover is 100%. However, there are very few real networks that are fan-out free. The single s-a fault cover gives an indication of multiple fault cover but no more. In the absence of reliable evidence of good cover, the reader should maintain a healthy scepticism.

It will be realised from reading Chapter 4 that a good test program generator will find tests for all faults for which a test exists. However, the following points were noted.

- Finding such tests is a very long and expensive process.
- Hand generated tests produced for design verification have been found to give quite high fault cover and the intelligence of the designer can produce more tests relatively easily and quickly. This leaves only the less obvious tests to be generated automatically. Reduction in the number of tests is important since the cost of testing is high – 60–70% of the total cost of a chip has been claimed.

The problem then is to find which faults are covered by the test set produced so far. Thus one can find the faults for which the automatic test program generator must try to find tests. Once a new test vector to find one of these faults is produced, it is necessary to see which other previously untested faults are also covered by that vector.

To determine which faults a test program can find, each fault is introduced into the network one at a time. A simulation is run against the test program and the results compared with the result of simulating a fault

---

[1]   B. Prior, Plessey Research (Caswell). Verbal statement at 'Systems on Silicon,' 1987.

free system. Thus, for this purpose, the design must have been accepted as satisfying the specification by the use of designer-generated tests on the simulator. The purpose here is to determine if a fault introduced in the *manufacturing* process (as distinct from a *design* fault) can be detected.

The cost of running a simulation with every individual fault is also very high. It is done only once and not every time a real system is tested. Nevertheless, it is very necessary to reduce the cost. For a 1M node circuit and considering only single s-a faults, 2M simulations are needed. The purpose of this chapter is to describe methods of doing this. Whilst considerable progress has been made, the continually increasing size of chips and systems is such that better methods are always needed. Current development would seem to be some way behind the desirable objective.

9.2   Reducing the problem size

Nishida (1987) describes the size of the fault simulation problem as a three-dimensional structure as shown in Fig. 9.1. The axes are the number of gates, G, the number of faults, F and the number of tests, T. Since both the number of faults and number of tests are in some way related to the number of gates, the size of the problem is, in fact, proportional to G^3.

Nishida suggests that the problem can be reduced by considering each of the three factors in both static and dynamic forms. By 'static' is meant things which can be done prior to simulation and hence independent of the progress of the simulation. By 'dynamic' is meant things that may be done as a result of observing the progress of the fault simulation.

Fig. 9.1. Diagrammatic representation of the size of the fault testing problem (Nishida).

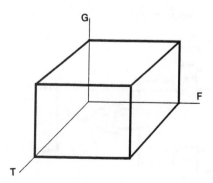

9.2.1 Static reduction of tests

The size of the 'gate' dimension can be reduced in two ways.

- Use of higher level elements. For example, if it is possible to work with an 8-bit ALU rather than with individual AND and OR gates, then the number of single s-a faults can be reduced from around 1200 to about 70 for this ALU.
- Removal of 'meaningless' gates. By this is meant inverters and buffers which have no *logical* function. For a buffer, for example, the function is 'copy.' It is impossible to distinguish between a s-a-*0* on the input and a s-a-*0* on the output. Thus, whilst some delay is still necessary, it is possible to remove two faults (s-a-*0* and s-a-*1* on one of the nodes) from the list that must be detected.

The above procedures inevitably reduce the number of faults to be simulated and hence the *F* dimension of Nishida's cube. A further obvious reduction can be made by fault collapsing. In a sense the removal of 'meaningless' gates could be included in this class. Other examples have been quoted in Chapter 4. For example, one *or more* s-a-*0*s on the input of a NAND gate is equivalent to a s-a-*1* on the output. Hence it is only necessary to test for one of these faults. For a four-input NAND gate, this is a reduction from five single faults to one. Furthermore, 10 double, 10 triple, five quadruple and one quintuple s-a-*0* faults are also covered, together with multiple faults involving s-a-*1*s at the input when at least one other input is s-a-*0*. The reduction in the number of faults involving s-a-*1* faults at the input and the s-a-*0* at the output is smaller, but is still significant.

A second method of reducing the number of faults for which tests must be made makes use of the properties of so-called fan-out free networks. Fig. 9.2 shows a fragment of a network. The term **fan-out free** implies that the

Fig. 9.2. Fan-out free region of a network.

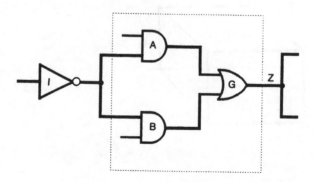

signals have one and only one load. The elements I and G in the figure have two loads each, but elements A and B have only one. The section of the circuit in the box is said to be a fan-out free region. It has been found that if a fault on the signal Z is undetectable then faults within the preceding fan-out free region are also undetectable. The converse may or may not be true. Further, many faults of the fan-out free region, including those on its inputs, can be replaced by a 'surrogate' fault at Z. No test can distinguish faults internal to the fan-out free network from each other or from the surrogate fault(s). Thus attempting to detect these faults is a waste of resources. This is an extension of fault collapsing.

It is also useful to note that, within the fan-out free region, critical path tracing (Section 9.7.1) often gives considerable speed-up over other methods of assessment.

A further method of reducing the F dimension of Nishida's cube is by use of parallel algorithms. This is the major subject of this chapter and further discussion is postponed.

On the test dimension, the number of actual tests can be reduced by fault merging. This was discussed in Section 4.5.

9.2.2 Dynamic reduction of the Nishida cube

The reader will by now be aware that the event driven simulation algorithm gives a reduction in the number of gates to be processed relative to the compiled code algorithm. It is mentioned here for completeness in relation to Nishida's cube. However, it is also noticed that for most faults, and always for single faults, the number of elements affected by a fault is small. If some method can be devised by which only this small number of 'gates' is simulated then the amount of work to be done in the fault simulator is reduced. This is the basis of the concurrent algorithm in particular (Section 9.4).

Considering the T dimension, suppose that the test program is 1000 vectors long. Suppose that a particular fault is found by the fifth vector. There is no point in running the other 995 tests on this fault. This fault is **dropped** from the list of faults for which checks are being made.

This procedure can be further improved by ordering the test vectors statically in such a way that those finding the most faults occur first. Sarfert et al. (1992) suggest that inserting faults near the primary inputs will find more faults than those inserted later in the network, which seems very reasonable. These should be simulated first. This means that many faults can be dropped early, leaving fewer to be sought. In a multipass program – which will be essential for large systems – this will significantly reduce the amount of work to be done. Both these techniques also have an effect on

Nishida's F dimension.

There is one proviso here. Hughes (1988) suggests that, if a single fault, P, is detected only once, there is a significantly greater chance that multiple faults including P will *not* be detected. On this basis, it may be worth *not* dropping the fault until it has been detected k times where k is some small integer. Indeed, this facility is available in a number of proposed systems.

Goel *et al.* (1980) suggest another method of reducing the number of faults to be considered with a given test vector. If a single fault is found not to propagate and hence is not tested, the fault is replaced by an X value and the X is propagated. Any signal which is changed to an X is not observable with this test vector and can be removed from the list of those for which checks are made.

Apart from the initial hand-generated tests for logic verification, tests will be produced automatically. This is done by selecting a fault and generating a test to detect it. However,

- this test may also find other faults, especially in a multi-output network,
- some 'don't care' inputs could be set to 'do care' values in order to find more faults with each single test (fault merging).

Hence, after each test or small group of tests is generated, that (set of) test(s) should be fault simulated. This then reduces the T dimension of Nishida's cube.

A further suggestion was made by Demba *et al.* (1990). In some cases it may be worth performing some sort of initialisation after a few tests. This brings the system state back to a 'standard' point. From here it may be easier to progress to detecting more faults than by simply trying to move forward from the current set of inputs.

The following sections describe several proposed methods of fault simulation which run faster than running one fault at a time on a conventional logic simulator. These will be compared, and a number of variations and enhancements discussed. Unless otherwise stated, single faults are assumed. The reader should consult the literature for discussion of multiple faults, notably Hughes (1988) and also Agarwal and Fung (1981). Much of the work in that area assumes fan-out free networks, which is rather far from real networks.

9.3   Parallel methods of fault simulation

There are two parallel approaches to fault simulation. In the 'older' method, a number of faults are simulated in parallel with a single test pattern. More recently, an approach based on simulating many patterns in parallel for one fault has been shown to be advantageous. The older method is often referred to in the literature simply as 'parallel fault simulation.' To keep things clear this text will use the terms 'parallel fault' and 'parallel pattern' respectively.

9.3.1 Single fault propagation

This method of fault simulation appears at first sight to be rather slow. However, combined with parallel operations, it is thought by many workers to be one of the best methods. In particular, it has low memory requirements. The explanation is done in conjunction with the not equivalence circuit reproduced as Fig. 9.3. To further simplify the explanation, the faults assumed will be limited to those due to faults in the circuit elements – that is, those on the element outputs. The discussion will be extended to faults in the wiring once the principles have been explained. The results for the input pattern $A = B = 0$ are shown in Table 9.1. In order to simplify the description, a serial single fault progapation procedure is explained first.

Firstly, simulate the good circuit. The results are shown in the first line of Table 9.1. Now introduce a fault and simulate until

- the fault is detected or
- the faulty circuit is the same as the good circuit.

In Table 9.1, the first fault introduced is A s-a-0, written for brevity as A_0. A being 0 is the same as the good circuit, so the faulty circuit will be the same as the good one and this fault is undetectable with the $A = B = 0$ test pattern. The second criterion is met.

Next introduce the A s-a-1 (A_1) fault. A is different to the good circuit so G1 and G2 are simulated. C remains the same as the good circuit, but D changes to 0 and hence Z to 1. As the circuit output is different to that of the

Fig. 9.3. Not equivalence gate example.

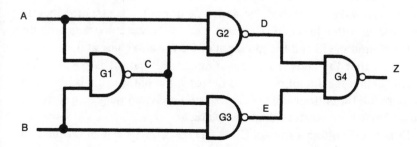

Table 9.1. *Single fault propagation on Fig. 9.3*

	A	B	C	D	E	Z	Faults found
Good	0	0	1	1	1	0	
A_0	0						
A_1	1	0	1	0	1	1	A_1, D_0, Z_1
B_0	0	0					
B_1	0	1	1	1	0	1	B_1, E_0, Z_1
C_0	0	0	0	1	1		
C_1	0	0	1				
D_1	0	0	1	1			
E_1	0	0	1	1	1		

good circuit, the fault A_1 is detectable – first criterion.

Comparing the third line of Table 9.1 with the first line (the good circuit), there are differences at D and Z as well as at A. Hence the faults D_0 and Z_1 can also be detected by this test. It will not be necessary to insert these two faults and fault simulate. The reader may like to check by inserting one of these faults.

Proceed in a similar manner with other faults. B_0 is not simulated (same as the good circuit). B_1 leads to a difference in Z so the fault is detected. Comparison of lines 5 and 1 of Table 9.1 shows that E_0 and Z_1 are also detected. Insertion of faults C_0, C_1, D_1, E_1 and Z_0 shows that these faults cannot be detected with this test. Note that D_0, E_0 and Z_1 were not inserted as they had already been detected.

The same procedure can be followed with the other three input patterns to the circuit. The reader can try this and will find the results below (Table 9.3). However, it is not necessary to simulate the faults already detected unless it is required to have a test program in which there is more than one way of detecting a particular fault.

9.3.2   Extension to include faults in the wiring

As indicated earlier, the above description has assumed that all faults occur within the circuit elements. Thus if C is stuck at 0 then both the bottom input of G2 and the top input of G3 are also stuck at 0. Suppose, however, that the fault is in the wiring. In this case it is possible (for example) that the output of G1 and the top input of G3 could be the good values while the bottom input of G2 is s-a-0. This is a consequence of signals being fanned out to more than one place.

Denote the bottom signal on G2 as CD (i.e. C on its way to D) and that

Table 9.2. *Extension to Table 9.1 for wiring faults*

	AC	AD	BC	BE	CD	CE	D	E	Z	Faults found
Good	0	0	0	0	1	1	1	1	0	
AC_0	0	0								
AD_0	0	0								
AC_1	1	0	0	0	1	1	1	1		
AD_1	0	1	0	0	1	1	0	1	1	AD_1, D_0, Z_1
BC_0			0							
BE_0				0						
BC_1			1	0	1	1				
BE_1			0	1	1	1	1	0	1	BE_1, E_0, Z_1
CD_0	0	0	0	0	0	1	1	1		
CE_0	0	0	0	0	1	0	1	1		
CD_1	0	0	0	0	1	1				
CE_1	0	0	0	0	1	1				

on the top input of G3 as CE, with corresponding names for other signals. There is no need to split signals D or E, which have only one load each. Single fault propagation can now be applied to the circuit with all 24 faults. Table 9.2 is an extension of Table 9.1 to handle the *extra* faults. The number of columns is extended to include the extra distinct signals. The faults AD_1 and BE_1 are detected with the *0 0* test vector. Application of all four possible input vectors shows that all 24 faults can be detected, but all four patterns really are needed. Remember, both Tables 9.1 and 9.2 are required for the full test.

9.3.3 Assessment

It is interesting to note the total number of runs required. Consider the $A = B = 0$ test vector, and assume that it is only required to detect each fault once. There is one good circuit and 24 faulty circuits. However, inserting the fault A_1 finds the faults D_0 and Z_1, and inserting the fault B_1 also finds the fault E_0. There is no point in searching for faults in which the faulty signal is identical to that in the good circuit. Thus the faults $A_0, AC_0,$ $AD_0, B_0, BC_0, BE_0, C_1, CD_1, CE_1, D_1, E_1$ and Z_0 need not be inserted. Hence the total number of simulations with this pattern is only 10. Seven faults are found.

Let the next test vector be $A = 1$, $B = 0$. There are 17 faults to be inserted, but it is possible for three of these, D_1, Z_0 and C_0, to be found along with other faults. There are seven new faults in which the faulty value is identical to that of the good circuit, three of which are in the previous category. A total of six runs, including the 'good' circuit, will find seven new faults. For the $0\ 1$ pattern, six runs will find five more faults and for the last vector, five runs will find the last five faults. A total of 27 runs is required, including those for the good circuit. The faults found by each pattern are summarised below. Those in brackets are also found in an earlier test.

$$A = 0, B = 0: \quad A_1, AD_1, B_1, BE_1, D_0, E_0, Z_1$$
$$A = 1, B = 0: \quad A_0, AD_0, (B_1), BC_1, C_0, CD_0, D_1, Z_0$$
$$A = 0, B = 1: \quad (A_1), AC_1, B_0, BE_0, (C_0), CE_0, E_1, (Z_0)$$
$$A = 1, B = 1: \quad (A_0, B_0), AC_0, BC_0, C_1, CD_1, CE_1, (D_0), (E_0), (Z_1)$$

Note that, if a run is done for each fault identical to a good output, the figure of 27 rises to 49.

For simplicity in what follows, the detailed description will usually be limited to the restricted set of faults. As an exercise, the reader should attempt to perform the fuller test. Results of such tests will be presented against which the exercise can be checked.

9.3.4 Parallel pattern single fault propagation (PPSFP)

The procedure just described is fairly fast and very economical on storage – only two values for each element input need to be stored at any one time. The process can be made faster by using parallel simulation.

Parallel simulation makes use of the facility, available in most computer instruction sets, of being able to perform logical operations between corresponding bits of a word. Thus, for a two-input NAND gate (e.g. G1 in Fig. 9.3), a four-bit word can be used to simulate all possible input patterns in parallel, thus

```
A   0 1 0 1
B   0 0 1 1
C   1 1 1 0
```

Consider, now Table 9.3. This shows the result of simulation of all four input patterns throughout the circuit of Fig. 9.3. Within each word, each bit is the value at that point in the circuit as a result of the setting of the equivalent bit of preceding signals. For example, take the second bit in the word. $A = 1$, and $B = 0$. This results, after simulation, in $C = 1$, $D = 0$, $E = 1$ and $Z = 1$, which can be seen at the second bit of the appropriate word.

Following the notion of *single* fault propagation the good circuit is first simulated. A fault is introduced. In Table 9.3 the first one is A_0. With this fault all four inputs for A must be 0. Simulation is repeated. Examining Z, it is seen that the second and fourth bits are different from the result of simulating the good circuit. Hence A s-a-0 is detected twice by this set of

Table 9.3. *Parallel pattern fault simulation (PPSFP)*

	A	B	C	D	E	Z	Faults detected
Good	0101	0011	1110	1011	1101	0110	
A_0	0000	0011	1111	1111	1100	0011	$A_0, C_1, D_1, E_0, Z_0, Z_1$
A_1	1111	0011	1100	0011	1111	1100	$A_1, C_0, D_0, E_1, Z_0, Z_1$
B_0	0101	0000	1111	1010	1111	0101	$B_0, C_1, D_0, E_1, Z_0, Z_1$
B_1	0101	1111	1010	1111	0101	1010	$B_1, C_0, D_1, E_0, Z_0, Z_1$

A	B	A_0	A_1	B_0	B_1
0	0		A_1, D_0, Z_1		B_1, E_0, Z_1
1	0	A_0, D_1, Z_0			B_1, C_0, D_1, Z_0
0	1		A_1, C_0, E_1, Z_0	B_0, E_1, Z_0	
1	1	A_0, C_1, E_0, Z_1		B_0, C_1, D_0, Z_1	

input patterns. Further comparisons of the first two lines of Table 9.3 show that other faults can also be detected. For example, the second bit of Z is *0* for the faulty circuit and *1* for the good circuit. Thus Z_0 can be detected. Bit 4 shows that Z_1 can be detected. Consider, further, C. The fourth bit is *1* in the faulty circuit and *0* in the good circuit. As the fourth bit of Z is different from the fourth bit of the good circuit the fault C_1 is also detectable. This could be confirmed by inserting the fault C_1 (C is all *1*s) and simulating (with A = *0101*).

Table 9.3 indicates all the faults detected. The second half of Table 9.3 extracts from the top half the faults detected by each input pattern with each of the single faults on A and B for comparison with Table 9.1 and equivalent tables for other patterns and with Table 9.6. Take the left hand bit of each word in Table 9.3 (A = B = *0*). Note that Z is different to the fault free value in rows A_1 and B_1. Consider only these two rows to construct the first row of the second half of the table. It is seen that A_1 is different from *0* in this bit in row A_1. So is B_1 in column B of the B_1 row. In column D, the first bit is different to the good circuit in the A_1 test, and hence D_0 is detectable. Thus is this row built up. The remaining rows of the second half of the table are built up by considering the other bit positions of the words.

To find test patterns to detect all s-a faults on the two primary inputs, obviously all four faults must be inserted. Observing the list of faults found, it will be seen that all other single s-a output faults within this circuit are also found. This is hardly surprising for such a small circuit. Indeed, in this example it will be seen that all faults on C, D and E are detected twice, and faults on Z are detected four times.

9.3.5   Extension to wiring faults

In considering an extension to the wiring faults, a little thought will show that each of the faults AD_v, BE_v etc. (where v is 0 or 1) will have to be separately inserted. The A, B and C columns of Table 9.3 will each be split into two, AC, AD; BC, BE; and CD, CE. It is not necessary to have an A, B or C column as well. To find all 24 faults, 17 runs are required, including that for the good circuit. As logical operations are the same speed, this represents a real speed-up. Table 9.4 gives the faults found for each of the introduced faults. It is not necessary to introduce separate faults on C, D, E or Z as these are already detected during simulation of other faults.

9.3.6   Evaluation

In a larger circuit there would be more test patterns. If there are W bits in a machine word then W patterns can be simulated in a single pass with one word per signal. This gives a speed-up of W in terms of the number of patterns simulated, but it will be appreciated that some work may be 'wasted' in detecting some faults more times than is really required. The reduction in the number of runs seen in the figures here is likely to be distorted by the very small size of the circuit. There were 27 runs for the single pattern simulation. Since all four patterns are run in parallel in the PPSFP method, the figure of 49 may be relevant. The figure of 27 shows how one may take advantage of a particular circumstance. The reduction from 49 to 27 may not be so dramatic in large circuits, but the drop to 17 with the parallel algorithm is likely to be much bigger in larger circuits.

As evaluation uses the parallel logical functions of the computer, multiple input gates must be simulated by a recursive process taking a new input with the result of the previous logical operation(s).

This method of fault simulation is primarily of use for simple gates, in order to use the word-wide logical operations of the computer. It could be used with more complex models, but there are two problems.

- Complex models do *not* use simple logical functions as a general rule. Thus they cannot make use of the word-wide logical functions. Consequently the individual bits for each test pattern must be extracted.
- Because the word-wide functions cannot be used, the model must be run W times in series. The advantage of parallel operation is lost – indeed, more than lost, since individual bits must be extracted from words. Thus the main usefulness of the technique is in IC design where gate level operations are likely to be used.

As described, the deductions made at various places depend on static

Table 9.4. *Faults found for each introduced fault*

Fault	Faults found	Test pattern(s)	
A_0	$A_0, D_1, Z_0, C_1, E_0, Z_1$	*1 0*	*1 1*
AC_0	AC_0, C_1, D_0, E_0, Z_1	*1 1*	
AD_0	AD_0, D_1, Z_0	*1 0*	
A_1	$A_1, D_0, Z_1, C_0, E_1, Z_0$	*0 0*	*0 1*
AC_1	AC_1, C_0, E_1, Z_0	*0 1*	
AD_1	AD_1, D_0, Z_1	*0 0*	
B_0	$B_0, E_1, Z_0, C_1, D_0, Z_1$	*0 1*	*1 1*
BC_0	BC_0, C_1, D_0, E_0, Z_1	*1 1*	
BE_0	BE_0, E_1, Z_0	*0 1*	
B_1	$B_1, E_0, Z_1, C_0, D_1, Z_0$	*0 0*	*1 0*
BC_1	BC_1, C_0, D_1, Z_0	*1 0*	
BE_1	BE_1, E_0, Z_1	*0 0*	
CD_0	CD_0, D_1, Z_0	*1 0*	
CE_0	CE_0, E_1, Z_0	*0 1*	
CD_1	CD_1, D_0, Z_1	*1 1*	
CE_1	CE_1, E_0, Z_1	*1 1*	

values. Thus, if the method is used with accurate timing, the check against the good circuit should be performed only after transient effects have died away, since differences in *1* to *0* and *0* to *1* delays may lead to spurious differences. Hence the event driven algorithm is not needed and there may be real advantage in using a compiled code technique. Finding faults due to timing problems is a task of a different nature – see Section 9.8.

9.3.7 Multiple values

It is possible to introduce multiple values into this type of simulation. For example, a three-value system would use two words per test pattern instead of one. By careful coding of the signal values, the number of logical operations between words can be minimised. Thus, if

$$0 = 0\ 0, \qquad 1 = 1\ 1, \qquad X = 0\ 1$$

the A AND B for all patterns is represented as in Table 9.5. In this table, corresponding bits of the two words are ANDed. It is seen that *0* AND *0 = 0*, *0* AND *X = 0*, *1* AND *X = X* etc. Finding similar codes for more values may be more difficult, but the principle remains valid. It will be appreciated that this approach is preferable to using $W/2$ patterns per word.

With multiple values, a problem arises if the good simulation gives a primary output of *1* or *0* and the fault simulation gives an *X* or vice versa.

Table 9.5. *Parallel evaluation of A AND B with three-value coding*

A	0	1	0	1	0(X)	0(X)	0	1	0(X)
	0	1	0	1	1	1	0	1	1
B	0	0	1	1	0	1	0(X)	0(X)	0(X)
	0	0	1	1	0	1	1	1	1
C	0	0	0	1	0	0(X)	0	0(X)	0(X)
	0	0	0	1	0	1	0	1	1

This is generally termed a **potential detection**, since whether the fault is detected or not depends on what the X value really is. Some additional inputs must be set, or some extra initialising done in order to get an actual detection (or not).

9.3.8 Parallel fault simulation

An alternative to the arrangement of the previous section is to regard the system with a fault as a separate network. There are then as many networks as there are faults. Given a two-value system, each word can hold the value for the good network and $W - 1$ faulty networks. Table 9.6 shows the example with all 24 possible faults being simulated in one pass. The faults are labelled along the top of the table. G is the good circuit. Only one test pattern is simulated, namely $A = B = 0$. The rows of Table 9.6 show the simulated values. It is not necessary to have a separate row for A, B and C (compare columns for an extended Table 9.3). In the first column, the good circuit, the values are as expected from Fig. 9.3. When a fault is introduced, differences occur. Consider the fault D_0. The fault does not affect A, B, C or E. D is forced to 0 regardless of the real value, which happens to be 1 (G column). As a consequence of this, the simulation finds that Z is 1. As the logic is the same for all networks, the word-wide logic of the computer can again be used for evaluation.

Now compare the values of Z in the faulty circuits with the value for the good circuit. It is seen that the values for the faults A_1, AD_1, B_1, BE_1, D_0, E_0 and Z_1 are all different, and hence these faults are detected by this pattern. None of the other faults causes a difference, so these faults are not detected. It can be seen that the results shown in Table 9.2 and in the second half of Table 9.3 are the same as those shown in Table 9.6.

For this simple circuit, all 24 faults of the full set can be simulated in one 32-bit word. In general this will not be true. If some but not all faults were detected with the first test, as is the case above, then in future tests it is necessary to perform the comparison for those faults so far undetected. A suitable mask will enable this to be implemented. In the simple case illustrated, this does not matter since there are less than $W - 1$ faults. When the number of faults is much greater than W, the number of words to be simulated may well be reduced considerably.

To attempt to compare this method with the PPSFP method, suppose the word length was 4 bits – all that was needed for the PPSFP algorithm. All four input patterns are needed. One good and three faulty circuits are simulated in each run. For each pattern, all 24 faults are simulated, giving eight runs times four patterns, or 32 runs. This is twice as many as for PPSFP (17). Waicukauski *et al.* (1985) have shown that the number of CPU operations for parallel fault simulation is proportional to G^2 where G is the number of gates, whereas the number for PPSFP is proportional to G. The former value is probably pessimistic due to the use of fault dropping. Nevertheless, for large networks it is very significant. It is possible to combine parallel fault simulation with the concurrent algorithm to be described next, to give useful improvements, Section 9.5.

9.3.9 Fault dropping

As an example of how fault dropping can reduce the number of runs, note that seven detectable faults were found in the first test. The next test, $A = 1$, $B = 0$ say, need not look for these seven. The second pattern searches for 17 faults. With 4-bit words, this requires six runs. Seven new faults are found. For the third pattern four runs find five more faults. The final pattern has two runs. The total is 20 runs. It is likely that this is pessimistic due to the small number of faults found per run – see Section 9.3.1. The 20 runs compares with 17 for PPSFP.

If one also does not insert faults in which the value is the same as that for the good circuit, a further reduction is possible. Suppose the good circuit and the first three faults are simulated with the $0\,0$ pattern. Considering the results from the good circuit, some nine further faults can be eliminated from consideration. Three faults have been simulated (unnecessarily as it turns out) leaving a further $21 - 10$ to be checked. Thus only five runs are needed with this pattern. Proceeding in this way, it can be shown that a total of only 13 runs is required.

These comparisons should not be treated too seriously as this network is *very* small. It must also be appreciated that elimination of faults from the set is not done without some work. In a large system, it is likely that several patterns would be simulated before attempting fault dropping.

Table 9.6. *Parallel fault simulation of Fig. 9.3*

	G	A_0	AC_0	AD_0	A_1	AC_1	AD_1	B_0	BC_0	BE_0	B_1	BC_1	BE_1	C_0	CD_0	CE_0	C_1	CD_1	CE_1	D_0	D_1	E_0	E_1	Z_0	Z_1
AC	0	0	0	0	1	1	0	0	0	0	0	0	0	0	0	0	0	0	1	0	0	0	0	0	0
AD	0	0	0	1	1	0	1	0	0	0	0	0	0	0	0	0	0	0	0	0	0	0	0	0	0
BC	0	0	0	0	0	0	0	0	0	0	1	1	1	0	0	0	0	0	0	0	0	0	0	0	0
BE	0	0	0	0	0	0	0	0	0	0	1	1	1	0	0	0	0	0	0	0	0	0	0	0	0
CD	1	1	1	1	1	1	1	0	0	0	1	1	1	0	0	0	1	1	1	0	0	0	1	0	1
CE	1	1	1	1	1	1	1	0	0	0	1	1	1	0	0	1	1	1	1	1	1	1	1	1	1
D	1	1	1	0	1	0	1	0	1	1	1	1	1	1	1	0	1	1	0	1	1	1	1	1	1
E	1	1	1	1	1	1	1	0	1	1	1	1	1	1	1	1	1	1	1	1	1	0	0	1	1
Z	0	0	0	0	0	0	1	0	0	0	1	0	1	0	0	0	0	0	0	0	1	0	1	0	1

9.4 Concurrent fault simulation

9.4.1 General description

In parallel fault simulation, a word holds the value of a particular node in the fault free and several faulty circuits. In the majority of cases, the value for the faulty circuit will be the same as the fault free circuit as the fault is not near this node. To avoid this waste, it is possible to keep a list of faulty circuits which have a value at this node which is different to that of the fault free circuit. This fault list can be propagated through the network. The concurrent and deductive fault simulators (Section 9.6.3) both do this.

The concurrent fault simulator processes a good circuit and a set of faulty circuits at the same time. There are two types of events:

- an event in the good circuit,
- an event due to a faulty circuit. This may be an event of a type not met previously.

An event in the good circuit has an associated list of faulty circuits. Each of these faulty circuits has a signal which has a different value from that of the good circuit with fault free inputs.

The great advantage of the concurrent simulation is that it processes faulty circuits using the same element models as for the fault free one. This is equivalent to generating a whole series of pseudo-elements for each real element, one for each fault, and storing the data with each. Hence the amount of storage is very large, and will vary as the simulation progresses. As the progress of a simulation is not easily predictable, neither are the storage requirements. However, logical operations are generally fast. The crucial point is that, because the simulator uses the same models as the good circuit, high level models can be used. None of the alternative methods of fault simulation have this property. It implies that there is no need for gate level models for higher level elements and hence a great deal of storage and processing can be saved.

Consider the two-input gate of Fig. 9.4 with fault free inputs 0 and 1 and hence an output of 0. This appears as the fault free gate at the top left-hand side of the figure. Two faults are possible at the inputs (single faults only are considered):

A s-a-1 leading to $Z = 1$
B s-a-0 leading to $Z = 0$

The former has an output different to that of the good circuit and so the fault is propagated. The latter has the same output as the good circuit so there is no point in passing the fault on to the rest of the network. In addition to the two input faults in Fig. 9.4, the gate itself may be faulty. This can be expressed by one or more faults. What follows will be limited to faults at the output of an element. In this case, the only fault of interest is Z s-a-*1* since Z s-a-*0* is indistinguishable from the good circuit. Elimination of such obviously undetectable faults reduces the amount of work to be done very considerably and is essential with this procedure.

Concurrent fault simulation is often explained in terms of a set of faulty gates. These are those shown in Fig. 9.4. The gate for B s-a-*0* (bottom left) is shown merging with the good gate and will disappear from the fault lists for this and subsequent logic elements. The gate for A s-a-*1* is top right and the **fault injection** gate for Z s-a-*1* is bottom right. Each fault is stored as a flag and three boolean values as shown in Table 9.7. In this table, the flag is written as the fault name for ease of reading but will usually appear as a number, the **fault number**. It many also happen that the A s-a-*1* fault will be fault collapsed with Z s-a-*1* since they cannot be distinguished. The B s-a-*0* fault will be removed. In the simple view, the set of faults of Table 9.7 will be stored as a linked list, one list for each gate, and each entry containing all the data for the gate.

A saving in storage can be effected by holding only the information for each input. There is a separate list for each input and hence only one value per entry is required. The saving does not appear to be great in this simple example, but will be more so in a large system.

It will be noticed that, as the simulation progresses, new gates are added to the network (A, Z s-a-*1*, Fig. 9.4) and others removed from it (B s-a-*0* in Fig. 9.4). This can be handled by an event driven simulator but not by a compiled code simulator in which the network is fixed at compile time.

Fig. 9.4. A fault free gate and several faulty ones.

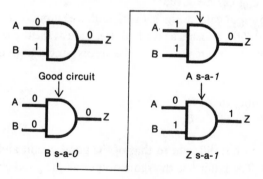

Table 9.7. *Storage of good circuit and fault list values*

Flag	A	B	Z
Good	0	1	0
A s-a-*1*	1	1	1
B s-a-*0*	0	0	0
Z s-a-*1*	0	1	1

9.4.2 Detailed example

To explain the processing, return to the not equivalent circuit redrawn as Fig. 9.5. The initial inputs are $A=0$, $B=0$. Results are shown in Table 9.8. Faults are labelled as 'X_v' for X s-a-v rather than using a fault number, which makes the explanation easier to follow. 'Ev' (event) is used to indicate when an event memory entry is to be made.

G1 is simulated first. The fault free output is $C=1$. The possible faults are A_1 and B_1 at the inputs and C_0 due to a fault in the gate itself. C_1 is the same as the fault free circuit and is not a possible fault with this test. It is not considered. Three extra gates are 'invented' but only two events are produced. These are the normal event for the fault free circuit and the fault event for C_0. The other two faults give the same output as the good circuit and so are undetectable with this test pattern. The associated gates cannot be deleted yet – see below.

Continuing the simulation resulting from the setting of A and B, fault free values of *1* appear at both D and E. In both cases the second input is not yet known since the value of C is a predicted value and time has not yet advanced to allow it to be set. There are four possible faults for both G2 and G3. For G2 these are A_1, C_0, C_1 and D_0. Notice again that D_1 gives the

Fig. 9.5. Not equivalence example.

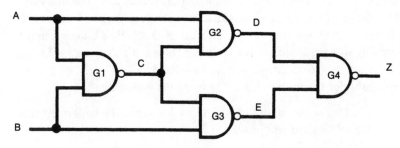

Table 9.8. *Simulation of Fig. 9.4;* $A = B = 0$

	C			D			E			Z		
	A	B	C	A	C	D	C	B	E	D	E	Z
Time 0												
Good	0	0	1 Ev	0	X	1 Ev	X	0	1 Ev			
A₁	1	0	1	1	X	X						
B₁	0	1	1				X	1	X			
C₀	0	0	0 Ev	0	0	1	0	0	1			
C₁				0	1	1	1	0	1			
D₀				0	X	0 Ev						
E₀							X	0	0 Ev			
Time 1												
Good				0	1	1	1	0	1	1	1	0
A₁				1	1	0 Ev						
B₁							1	1	0 Ev			
C₀				0	0	1	0	0	1			
D₀						0				0	1	1
E₀									0	1	0	1
Z₁										1	1	1
Time 2												
A₁										0	1	1
B₁										1	0	1

same value as the good circuit and is not considered. A_1 has an X as output since input C is unknown. Both C stuck faults lead to a *1* output which is indistinguishable from the good circuit (remember, only one fault is allowed). As the real value of C is not yet available, both faulty gates are retained at this time although no event is predicted. D_0 is a real fault and an event is produced.

G3 is treated similarly. There are not events on the inputs to Z at this time so it is not simulated. E_1 is a further fault which is not considered.

For simplicity, assume that all gate delays are the same and for convenience let them be unit time. Time now advances to 1. The first event found is the good event at C which causes no change to the good values of D and E. G4 is simulated with the good values of D and E as found at time 0 and leading to a prediction of the fault free value *0*. Fault gates for D_0, E_0 and Z_1 must be generated. Z_0 is not considered.

Next consider the fault events. These are the following.

- C_0. This causes both D and E to be *1* which is the fault free value. No new events are produced.

- D_0 and E_0. The values are different to the good values so G4 is simulated, giving $Z = 1$. This is different to the good value of Z, so an event is generated.
- A_1. At this time the gate of interest is G2, not G1. As C is now known to be 1, this fault leads to D being 0 which is different to the fault free value. An event is generated.
- B_1 has a similar effect on E, an event being generated.

Note also that the C_1 fault now gives the same inputs as the fault free circuit. This fault can now be removed from the fault lists of both G2 and G3.

Time now advances to 2. There are no fault free events, but there are two fault events. A_1 caused D to be 0 at time 1 and that fault propagates to Z as a 1. B_1 has a similar effect.

The simulation is now complete. The detectable faults are those in the list for G4 (Z) whose output value is different from the good value, namely, A_1, B_1, D_0, E_0 and Z_1. This will be seen to agree with what was determined earlier. Fig. 9.6 indicates diagrammatically the gates which have been generated in this simulation. In the simulator, G1 has three items in its fault list, A_1, B_1 and C_0. G2 and G3 also have three items. C_1 has been omitted as it occurred only due to the unknown initial value of C. G4 has five items as indicated above. It will be appreciated that the size of lists can become very large and hence the storage control is difficult.

Fig. 9.6. Circuit with fault list gates for Fig. 9.4 $A = B = 0$.

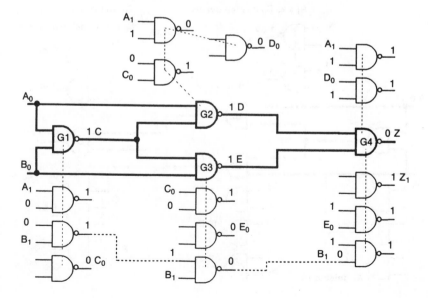

9.4.3   Change of input

Now simulate for an input change of A to *1*. For convenience let this happen at time 10. G1 and G2 are scheduled for the good circuit. Consider first G1. The good output does not change and no event is scheduled. However, it is necessary to simulate each of the faulty circuits of Table 9.8 and Fig. 9.6 which make up the fault list for G1. Table 9.9 and Fig. 9.7 illustrate what is happening.

- A_1. As A is now *1*, this is the same as the good circuit and this fault must be dropped. The gate in Fig. 9.6 disappears. However, observing Fig. 9.6 it is seen that gates labelled A_1 appear alongside G2 and G4. These must also be removed. To do so an event must be scheduled. This is a new type of event.
- B_1. C becomes *0*, which is different to the good circuit, in contrast to the state in Table 9.8. A fault event is scheduled.
- C_0. This is the same as for the previous set of inputs. A new event is not required.

G2 must also be simulated at this time. C was *1* from the previous step (time 1), so D becomes *0*. This is a change, so an event must be scheduled. The faulty gates of G2 are then simulated.

Fig. 9.7. Change of circuit due to input change. Compare Fig. 9.6.

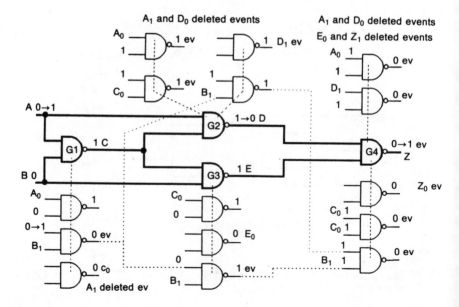

Table 9.9. *Change of input following Table 9.8*

	C			D			E			Z		
	A	B	C	A	C	D	C	B	E	D	E	Z
Time 10												
Good	1	0	1	1	1	0 Ev			(1)			
A₁	delete gate Ev			delete gate Ev								
B₁	1	1	0 Ev									
C₀			0	1	0	1 Ev	0	0	1			
A₀	0	0	1	0	1	1 Ev						
D₀				delete gate Ev								
D₁						1 Ev						
Time 11												
Good										0	1	1 Ev
A₁										delete gate Ev		
B₁				1	0	1 Ev	0	1	1			
C₀										1	1	0 Ev
A₀										1	1	0 Ev
D₀					0					delete gate Ev		
D₁										1	1	0 Ev
E₀							0			0	0	1
Z₀												0 Ev
Z₁										delete gate Ev		
Time 12												
B₁										1	1	0 Ev

- A₁ gives the same result as the good circuit and the fault gate must be deleted. An event is scheduled to enable the fault gate attached to G4 to be deleted.
- C₀ causes D to be 1. A fault event is scheduled. Note that E does not change.
- D₀ is the same as the good circuit. The fault gate is removed and an event scheduled to remove the gate attached to G4.

As A is now 1, a new fault, A₀, needs to be introduced. This gives C as 1, the same as the good circuit, but D becomes 1 which is different to the fault free value. As the good value of D has also changed, the D₁ fault must be introduced. It is noted in the previous paragraph that D₀ is in the process of being deleted.

At time 11 the good event on D is read. This causes G4 to evaluate and Z changes to 1. As a direct consequence of this, fault gate Z₁ must be deleted and fault gate Z_0 must be created. Both actions cause an event to be scheduled.

At time 10, events were generated deleting gates concerned with the A_1 and D_0 faults. These fault gates are also deleted with appropriate events for following logic. The other events are the following.

- B_1. As A is now 1, the B_1 fault caused C to become 0 at time 10, and that event results now in a fault gate on G2 with $D = 1$. An event is scheduled.
- C_0, A_0 and D_1 all cause D to be 1 which is different to the fault free value, and that in turn results in Z becoming 0. Events are scheduled in each case.

Note that the fault gate for E_0 is still present but has not been activated in this simulation cycle.

Finally, at time 12, the B_1 fault propagating via G2 reaches G4. Z is 0 and a fault event is generated.

Again the detectable faults are those giving values of Z different to the good value. These are A_0, B_1, C_0, D_1 and Z_0. As an exercise, the reader may like to repeat the process for the other two input patterns. The detectable faults are

$$A = 0, B = 1: \qquad A_1, B_0, C_0, E_1, Z_0$$
$$A = B = 1: \qquad A_0, B_0, C_1, D_0, E_0, Z_1$$

This algorithm can be applied with the additional faults due to individual wires stuck. Doing so will give results looking even more complex than those above and will add little to understanding. The reader may like to attempt this extension as an exercise. The results should be the same as those found in previous sections.

The data structures needed to implement the ideas described can be worked out fairly easily. d'Abreau and Thompson (1980) offer one possible set.

9.5   Parallel value list (PVL)

The problem with the concurrent method of fault simulation, as with the simple simulation, is the enormous store requirements. It has been seen that one method of alleviating this is to store the data for several fault machines in a single computer word, Section 9.3.8. This can also be used with the concurrent algorithm (Moorby 1983, Son 1985). A *group* of faults is carried around as a single unit. In the normal case two words – 64 bits – are needed for a single fault, assuming the linked list per signal variant (see last paragraph of Section 9.4.1). That is, 32 bits for a pointer and 32 bits for a fault number and a value. Only 2 bits out of 64 are used for real data (three-value).

Suppose instead that there are five words to hold 32 faults. Two words hold pointer and group number. A particular fault is identified by group number and position in the word. A third word holds a mask and the other two hold 32 values in the manner described in Section 9.3.8. Thus 64 bits out of 160, or about 13 bits out of 32, hold real data. The mask word is needed so that faults which have been detected can be prevented from affecting the progress of the simulation.

9.5.1 Detailed example

Consider the circuit of Fig. 9.8. Let the inputs both be 0, so that $C=D=E=1$ and $Z=0$. There are 24 possible faults – s-a-0 and s-a-1 on each node and fan-out node. Suppose that the computer has a 4-bit word and that the faults are divided into groups as follows.

Group 1 A_0 B_0 C_1 BE_1
Group 2 AC_0 BC_0 CD_1 CE_1
Group 3 AD_0 D_1 BE_0 E_1
Group 4 AC_1 CD_0 CE_0 BC_1
Group 5 D_0 E_0 Z_0 Z_1
Group 6 A_1 B_1 C_0 AD_1

The position in the lists is significant. Table 9.10 shows the initial state. As $A=0$ and D has not yet been reached, the only active faults at AC are A_1 and AC_1. A_1 affects AD as well, of course, but that is not relevant in relation to AC. The two faults are in the first position of group 6 and the first position of group 4 respectively. Clearly these are both 1. Groups 4 and 6 are attached to AC. The remaining positions in these two groups are set to

Fig. 9.8. Example circuit.

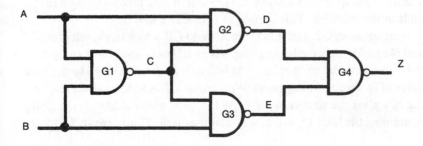

Table 9.10. *Parallel value lists* – $A = B = 0$

Signal	Good value						
AC	*0*	→	group 4 *1 0 0 0*	→	group 6 *1 0 0 0*		
AD	*0*	→	group 6 *1 0 0 1*				
BC	*0*	→	group 4 *0 0 0 1*	→	group 6 *0 1 0 0*		
BE	*0*	→	group 1 *0 0 0 1*	→	group 6 *0 1 0 0*		
CD	*1*	→	group 4 *1 0 1 1*	→	group 6 *1 1 0 1*		
CE	*1*	→	group 4 *1 1 0 1*	→	group 6 *1 1 0 1*		
D	*1*	→	group 5 *0 1 1 1*	→	group 6 *0 1 1 0*		
E	*1*	→	group 1 *1 1 1 0*	→	group 5 *1 0 1 1*	→	group 6 *1 0 1 1*
Z	*0*	→	group 1 *0 0 0 1*	→	group 5 *1 1 0 1*	→	group 6 *1 1 0 1*

the good value of the signal, namely AC, 0. Node AD has A_1 and AD_1 active in group 6 by a similar argument. Nodes BC and BE are assigned fault groups in a similar manner. It is not necessary to consider nodes A and B separately.

Consider now node C. As the function of G1 is NAND, the bits of the groups associated with AC and BC are considered. Both inputs have groups 4 and 6 attached. Consider group 4 first. The corresponding bits are NANDed. All four bits are *1* which is the same as the good value of C. Group 4 disappears. However, the active faults CD_0 and CE_0 also happen to be in group 4, so on CD this group reappears with the second bit at *0* and on CE with the third bit at *0*. Group 6 is treated similarly. Again, all bits are *1* and the group should disappear. However, it is reintroduced with the C_0 fault in the third bit. This applies to *both* CD and CE.

Next consider G2. The inputs are AD and CD which have fault groups 6 and (4 and 6) respectively. As group 4 is on CD only, the bits of group 4 are NANDed with the good value of AD (*0*) giving all values equal to the good value of D. Group 4 disappears. With group 6, NANDing the corresponding bits gives the pattern *0 1 1 0*. As the good value of D is *1*, the group containing the fault D_0 must be attached as well. This is group 5 with the first bit set to *0* and the remainder to *1*. The interpretation of this line of the table is that faults A_1 and AD_1 will both cause faulty values of D.

G3 has inputs BE and CE. These have fault groups (1 and 6) and (4 and 6) respectively. Group 1 is on BE only, and so NANDs with the good value of CE (*1*) to give an inversion. Group 4 NANDs with the good value of BE (*0*) to become the same as the good value of E and hence inactive. The two group 6s NAND together giving a *0* in the second bit and *1*s in the remainder. Group 5 is added with the fault E_0 set to *0* and the remaining bits to *1*, the good value of E.

Finally, for G4, the fault groups 5 and 6 are on both D and E and group 1 is on E. As the good values of both D and E are *1*, group 1 copies to Z with the bits inverted. Group 6 transfers as shown. Group 5s NAND to give *1 1 0 0*. However, the fault Z_1 has to be added as a *1* in the last bit.

Comparing the good values of Z (*0*) with the values in the fault groups for Z, it is seen that seven places are different. Comparing these with the group definitions, it is seen that the faults BE_1, (group 1) D_0, E_0 and Z_1 (group 5 in order) and A_1, B_1 and AD_1 (group 6 in order) are detected. Comparison with previous results will show that this is the same as was obtained by other methods.

9.5.2 Change of input

Suppose now that A becomes *1*. A_1 and AC_1 are no longer active faults but A_0, AC_0 and AD_0 are. Hence the fault groups for AC and AD change and events are created. B has not changed, so the fault groups for BC and BE are unchanged and no events are generated. Table 9.11 shows the new lists.

As the fault group for A changed, the faults for C and D must be re-evaluated. For G1, there are no fault groups on AC which also appear on BC and vice versa. The good value of B is *0*, which, when NANDed with the values in the AC fault groups (1 and 2), leaves the groups with all values equal to the good value on CD and CE and hence inactive. However, groups 4 and 6 on BC NANDed with the good value of A (*1*) puts groups 4 and 6 on CD and CE. The C_0, CD_0 and CE_0 faults must be added in the appropriate positions as before.

For G2, a fault on either input can now cause an output fault. Thus groups 1, 3, 4 and 6 all pass to the fault list of D with an inversion of values. D_1 is added to group 3 and D_0 removed as inactive, which removes group 5 completely (compare with Table 9.10).

With G3, BE and CE both have fault group 6. NANDing the bits together gives all *1*s, and the group becomes inactive and disappears. Group 1 from BE with the good value of CE (*1*) produces an inversion of the

Table 9.11. *Parallel value lists*: $A = 1$, $B = 0$

Signal	Good value						
AC	1	→	group 1	→	group 2		
			0 1 1 1		0 1 1 1		
AD	1	→	group 1	→	group 3		
			0 1 1 1		0 1 1 1		
BC	0	→	group 4	→	group 6		
			0 0 0 1		0 1 0 0		
BE	0	→	group 1	→	group 6		
			0 0 0 1		0 1 0 0		
CD	1	→	group 4	→	group 6		
			1 0 1 0		1 0 0 1		
CE	1	→	group 4	→	group 6		
			1 1 0 0		1 0 0 1		
D	0	→	group 1	→	group 3	→	group 4
			1 0 0 0		1 1 0 0		0 1 0 1
				→	group 6		
					0 1 1 0		
E	1	→	group 1	→	group 5		
			1 1 1 0		1 0 1 1		
Z	1	→	group 1	→	group 3	→	group 4
			0 1 1 1		0 0 1 1		1 0 1 0
				→	group 5	→	group 6
					1 1 0 1		1 0 0 1

bits. Group 4 from CE with the good value of BE (*0*) becomes inactive and disappears. Group 5 is added with E_0 set at *0* and the remaining bits at the good value of E, namely *1*.

Finally, for G4, groups 3, 4 and 6 from D are NANDed with the good value of E (*1*) and group 5 with the good value of D (*0*). This removes group 5 but it is then replaced with the fault Z_0 included as *0* at the third bit and the rest of the bits as the good value of Z, *1*. Group 1 is common to D and E so the bits are NANDed.

By comparison of the group values with the good value of Z, it is seen that the faults detected are A_0, AD_0, D_1, CD_0, BC_1, Z_0, B_1 and C_0 in the order in Table 9.11. Again, comparison with previous results will show that the same conclusions have been reached.

9.5.3 Comment

The problem with this sort of representation is that either it must be restricted to simple gates or the data has to be extracted separately for each fault circuit. This means that the high level capability of the concurrent algorithm is lost. On the other hand, the algorithm is still of use in chip design where a low level of representation is essential at some stage.

A second problem is that the efficiency of operation is very dependent on the way the faults are grouped. Table 9.10 has a total of 19 groups attached to various signals. A 'less intelligent' attempt at choosing groups tried in the course of preparing this text gave 22. Efforts should be made to group faults which are likely to occur together in order to keep the fault activity high and the number of active groups low. How that should be done is not easy to determine.

9.6   Assessment of simulation methods

9.6.1   Parallel methods of fault simulation

The parallel methods of fault simulation are fundamentally tuned to the word size of the machine on which the simulation is run. This is probably 32 bits, but may be already 64, and certainly will be within a few years. For simple gates – AND, OR, NOT and possibly not equivalence – the speed-up is $W-1$, where W is the word length, since, in reality, single bit logical operations usually take place on full words. Thus a word length operation is the same speed as a bit operation. A considerable saving in storage is certain. Current evidence suggests that parallel pattern methods are faster than parallel fault methods.

Multivalue signals can be handled. Some authors have suggested that more than three are useful, but fault simulators are not the best places to look for timing errors (see Section 9.8). Multiple word manipulations need be no more difficult than multiple bit ones for reasons just given, so, comparing like with like, it should be possible. Timing simulations are quite difficult, but again the event driven technique with full words per value should be possible.

Storage requirements are modest and predictable. Indeed, storage for $W-1$ faults may be the same as for one fault. It is possible to simulate all faults in one pass. There would seem to be little advantage in doing so, and a positive disadvantage in having to store all faulty machines. Multiple pass at $W-1$ faults per pass is no slower and is more economical in storage. Furthermore, by using fault dropping, the number of passes to be used can be significantly reduced dynamically.

The major disadvantage of parallel methods of fault simulation is that they cannot usefully be extended to models of a higher level than that of simple gates. This applies equally to the parallel value list algorithm with the concurrent one. Consider simulating a 4-bit ALU, Fig. 9.9. There are two operands of four bits each, a carry_in and five controls, a total of 14 inputs. A 14-input AND gate can be modelled by taking words for two inputs, and then successively ANDing the result with a new word. With the ALU, the result requires all inputs to be considered together and in a (very) complex function. The function can be evaluated sensibly only by using one set of data at a time. Thus one of 32 bits of data must be extracted from 14 different places and 32 evaluations performed one after the other. The speed advantage of parallel operation has been more than lost, though the parallel storage is still useful.

9.6.2   Concurrent fault simulation

In essence concurrent fault simulation simulates all faults together. In practice it is sensible to simulate only a limited number. There are two reasons for this.

- Store requirements vary unpredictably and widely due to expansion and contraction of the number of fault elements being simulated. Restricting the number of faults being simulated puts an upper bound on the maximum storage required.
- Combined with dynamic fault dropping, the total number of faults that must be simulated can be reduced, since some tests will detect several faults.

It is the unpredictability of the storage requirement that is the greatest

Fig. 9.9. An ALU.

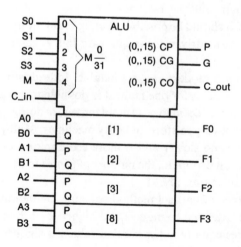

problem. On the other hand, 'store is cheap.' One doubts if this is ever really true, and as technology advances, so does the demand, the one never catching up with the other.

The major advantage of concurrent fault simulation is that it uses the logic models of the logic simulator. Thus is has the following properties.

- It is transparent to the logic values being used.
- It uses the same basic simulator. In particular, it uses the event driven approach so that only logical or fault events are processed.
- Because the event driven algorithm is used, accurate timing models can be used if really required (but see Section 9.8).
- High level models can be used, reducing the number of elements, the number of faults to be simulated and the maximum store requirement. Furthermore, hardware models can be used. When it is required to simulate an input s-a fault, the appropriate value is supplied to the hardware model. For an output s-a fault, the value read from the hardware model is ignored and replaced by the stuck value.

Rogers *et al.* (1987) take this further by including a program to combine several 'moderate level' modules into a higher level module. An example is described in which hierarchical simulation gives a speed-up of 30, with a further factor of six on combining modules by program.

It can be argued that the most important application of fault simulation is in ehip design where much of the design is at gate level. Hence techniques limited to gate level are not at a disadvantage. In testing PCB systems, it is possible to monitor internal points. This argument ignores the trend to design chips with higher level modules as well.

Concurrent fault simulation handles only changes from the good circuit. The area of influence of a given fault will usually be limited. On a qualitative basis, such a fault simulation ought to be fast. Of course, certain signals, such as resets or bus enables, which have high fan-out will be exceptions, but such signals are relatively few in number. It will be fairly obvious that concurrent simulation will be at its best when the fault activity is low, in contrast with parallel simulation.

Gai *et al.* (1987) have pointed out that the concurrent algorithm can be used with several similar input patterns on a good machine to accelerate normal simulation. It is conceivable that this should apply to parallel algorithms as well.

9.6.3 Deductive fault simulation

Like the concurrent fault simulator, the deductive simulator

(Armstrong 1972) holds lists of faults for each node, and these lists are propagated through the network. However, the method of propagating the lists is by means of set operations. These do not have the advantage of the operations of the parallel simulator. Nor can they make use of 'normal' models. Hence extension to multiple values or to general higher level models is difficult. Slightly higher level models have been proposed by using tables, but for more than a few inputs the tables become impractically large. Time dependent faults cannot be modelled.

For these reasons, deductive fault simulation has fallen out of favour but continues to be included in descriptions of available techniques. Thus it is mentioned here for completeness. Whether fresh developments could make the idea more viable is impossible to predict.

9.7   Some alternatives to fault simulation

Previous sections of this chapter have described basic approaches to fault simulation. The literature on the subject is very extensive. Space and time do not permit a full discussion of all the possibilities. This section reviews some of the ideas which appear to this author to be most promising.

9.7.1   Critical path tracing (Abramovici *et al.* 1984, 1990)

Critical path tracing is basically a test generation method which also gives a list of detectable faults. It is based on the notion of critical inputs to a gate. Consider a two-input AND gate as shown in Fig. 9.10. If one input is *1*, the output is a copy of the other input. This other input is the **critical input**. Hence in Fig. 9.10(*a*) A is critical. In Fig. 9.10(*b*) both inputs are critical.

With a given network, begin at the output. If the gate is an AND (or NAND) and the output is *1* (*0*) then both inputs are set to *1* and both are on the critical path (Fig. 9.10(*b*)). If the output is postulated as *0* (*1*) then one input is set to *1* and the other to *0* according to the values found by the good simulation, which has been done previously. The *0* is on the critical path. A comparable set of arguments can be applied to OR gates. The paths are traced back to the primary inputs using a D-algorithm style trace back.

Fig. 9.10. Critical inputs of an AND gate.

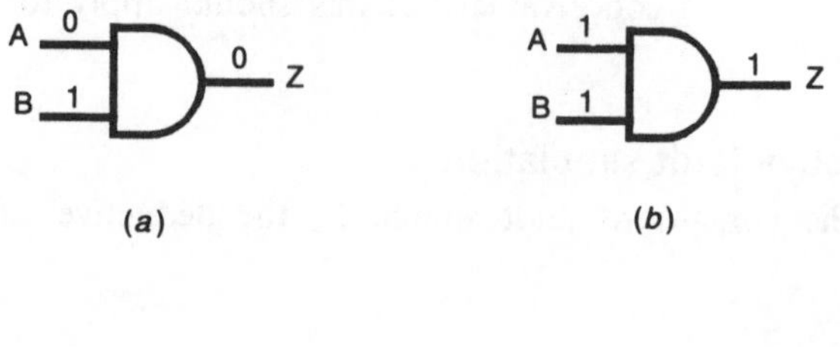

(*a*)   (*b*)

Where there are parallel critical paths, all are traced. Detectable faults are signals on the critical path(s) which are opposite to those of the good circuit. Thus a test vector has been generated together with a list of the faults it will detect.

The critical path trace algorithm only works well in fan-out free regions. Since few (no?) real networks are fan-out free, it is unlikely to be used alone. On the other hand, within the fan-out free region, it is very fast. Hence it may be used to advantage in combination with, for example, the PPSFP method. There is also a possibility that a parallel version of the critical path trace algorithm could be produced.

9.7.2 Statistical methods

A number of authors have suggested statistical methods of estimating the fault coverage of a set of test vectors. This might be based on information gathered during a simulation of a good circuit. For a given set of test vectors, the number of times a node is set to *1* and set to *0* will be an indication of the likelihood of detecting a s-a-*0* and s-a-*1* respectively.

Given the importance of high fault coverage indicated at the beginning of the chapter, statistical methods would seem highly suspect. Some example circuits tested by Jain and Agrawal (1984) suggested that it is possible to get within 5% of the real result. However, it is reasonable to expect there to be some pathological cases where the difference is much larger. Even 95% fault coverage is not good enough.

On the other hand, normal fault simulators only attempt to assess a limited set of fault models, usually single stuck at faults. Rogers *et al.* (1987) state

> Fault models on which fault coverage comparisons are based are only abstractions of the effects that possible physical failures could produce. Therefore computer fault coverages are predictions and exact measures are impossible.

A comment by Ulrich and Suetsugo (1986) is also apposite.

> The average simulation run is not successful. It produces incorrect results, too much useless information, too little useful information and wastes storage and CPU time.

Both comments suggest that statistical methods may, in practice, be no worse than supposedly more exact methods.

Following the above comment, Ulrich and Suetsugo propose a 're-hearsal' run with just a few faults to observe a limited number of points in the simulation. From the results of this run it is noted how much activity there is. If the activity is small then that particular part of the (fault) simulation is abandoned. Ulrich and Suetsugo claim that rehearsal runs:

- reduce CPU and memory requirements,
- rapidly determine if a test vector is inappropriate,
- provide a good estimate of resources needed for a full run,
- give an information gain per run as the end of the test set approaches.

9.7.3 Block orientated fault simulation

In Chapter 4 (Fig.4.18), a method of partitioning a system into blocks for test generation was suggested. Each block had a test program capable of finding 100% of faults in a given class. In a complete system, inputs to surrounding blocks were set in such a way as to transmit primary inputs to the block under test and to make the outputs of that block observable at the primary outputs. Freeman (1988) proposes a similar arrangement for fault simulation, largely for data path logic. He defines several types of block, but, in particular, an F-path block. This block is capable of having some set of inputs transferred to the outputs unchanged. To do so, it may be necessary to set the 'other inputs' to appropriate values. Consider the ALU shown in Fig. 9.9. If the carry_in is set to 0 and S is set to 'copy A' then F becomes A and the block is an F-path. It may also be an F-path if B = 0 and S is 'add' or 'subtract' or . . .. Because a system consists of rather few blocks the relevant F-path blocks can be found manually more easily than by program – people are more intelligent and better at pattern matching. The 'other input' values can also be specified manually at the same time. Freeman finds that he is able to get very compact tests for 100% single stuck at fault coverage, as well as for many other faults. As the method involves high level models, it is likely to be very fast. Whether it can handle non-data path logic is not yet clear.

9.8   Timing in fault simulation

9.8.1 Delay faults

The emphasis in this chapter so far has been on detecting stuck at faults. A class of fault which is critical, however, is timing faults. The question may be raised as to whether a fault simulator should look for these.

The problem is that, although the design may be acceptable after timing verification, a particular instance of a network may have a section which runs slowly due to manufacturing problems. If this results in the clock period being too short then the fault should be easily detectable.

There are two problems.

- To detect a slow signal path it is necessary to time a signal. To do this, *two* test vectors are needed. The first sets up a particular condition in the path or gate of interest. The second causes the transition of interest to pass along that path in such a way that the primary output on that path changes state. The delay of the path can then be measured.

- The fault may produce a short pulse. If this triggers some sequential circuit then the fault will be detectable, but what if the pulse is so short it will not trigger the sequential circuit reliably? What if spike removal in the simulation removes the spike when it should not? In some cases one may spend a lot of time looking for a logic fault when the problem is really in the timing. There are a great many difficulties in ensuring that *all* timing faults are found and this will depend on sequences of test vectors as much as on individual vectors.

Some writers have looked for individual gates that are slow. Rather more have been concerned about finding complete paths which are slow. In either case, full paths have to be traced out and controlling signals have to be set. To form a good test, all the controlling signals must reach their final value before the signal on the path reaches that point so that the longest path through the logic is on-path rather than some other path. Discussion of this topic uses the terms **robust** and **non-robust** tests. In any case, each test nominally requires a *pair* of vectors.

Notice that this discussion is concerned with testing the manufactured device for paths which are slower than worst case specification because of manufacturing problems. There is no doubt these faults must be found. If they reach the system they not only cause trouble. They are difficult to find. However, it seems to this writer that a different type of simulator is needed to find and verify tests for this purpose. Such simulators, known as delay fault simulators, are not yet well developed and the reader must look elsewhere for further details. A selection of references is included. They may or may not make use of fault simulator principles.

Although this uprates the parallel (and deductive) simulators relative to the concurrent one, the other advantages of the concurrent simulator are dominant, especially the high level model capability.

9.8.2 Oscillations and hyperactivity

A further problem may arise. Under fault conditions, feedback loops may be created, causing unscheduled oscillations to occur. This can cause an explosion of data in the fault simulator. It has been suggested that hyperactivity in an uninitiatable faulty machine can use up 90% of the time of a fault simulator. (Gai *et al.* 1988). It is useful if there is a mechanism which looks for 'hyperactivity.' When this is found, one approach is to assume that the hyperactivity will be such that the fault will be detectable. The simulation is purged and restarted without the offending fault.

Another circumstance which is sometimes used to abort a simulation is if too large a proportion of signals are in an X state after initialisation (Gai *et al.* 1988).

10

Simulator features and extensions

The earlier chapters of this book discussed the design and use of a simulator for use in the development of digital electronic systems. The discussion has been widened to include some aspects of testing and design for testability, since application of good practice in these areas leads to better use of costly resources in what is probably the largest part of the design procedure. It is now of value to review the extent to which the aims of simulation can be achieved; to discuss several topics related to the use of simulator; to introduce some enhancements to simulators; and to attempt to look into the future.

10.1 Desirable features of a simulator

Some years ago the author wrote down a list of the features he would like to find in a simulator.

1. A simulator is required to give an accurate prediction of the behaviour of a *good* network.
2. A simulator is required to recognise and give warning of a faulty network.
3. The basic simulator should be independent of technology but recognise the distinctive features of known technologies. Thus devices of *any* technology might be simulated.
4. The simulator should be capable of handling models at several levels of abstraction and in the same run (Harding 1989).
5. There is no point in simulating a design in 1 s if it takes a day to diagnose a fault, modify and recompile the network. Hence, associated with the simulator, there must be means to assist the user to find the source of 'wrong' results, correct them and

recompile quickly. That is, the simulation *cycle* must be given serious attention (the detail is not within the scope of this book).

The author also added the following comments.

6. When a simulator finds a timing error, there is no way it can know what the real logic would do. Should the simulator

    *   set an unknown state?
    *   estimate the designer's intention and carry on?
    *   mirror as faithfully as possible what the real logic would do?

7. There is no point in designing a 10M gate chip which works perfectly to specification *if the specification is wrong* (Harding 1989, Hodge 1990).

8. It takes as long to design a good model for an element as to design the element itself (other than very simple elements, perhaps). Hence hardware modelling for larger elements is essential.

9. 'Spike removal' by use of inertial delays will remove a spike that will cause an error in real hardware.

    Reports of all spikes will create such a mass of output data that the designer will either ignore it or go mad in the attempt to read it.

    Therefore some intelligent compromise is essential.

Abramovici *et al.* (1990) also set down some desirable features of a simulator. These are complementary to those above.

(a) The simulator should verify correctly, independent of the power on state.

(b) The simulator should be independent of component delays.

(c) The simulator should not be subject to critical races, oscillations, hang-up states etc.

(d) It should be possible to play 'what if?' games rapidly.

(e) It should be possible to evaluate design changes rapidly.

(f) It should be possible to produce timing diagrams for documentation.

In comparison with hardware, prototype or production, a simulator has the following abilities.

(g) It can check error conditions.

(h) It can check worst case timing.

(i) It can start in any desired state (thus meeting feature (a) above).

(j) It can give precise control of timing so that details can be checked.

(k) It enables debugging of software and ROM code before hardware is constructed.

(l) It is possible to 'probe' inside a piece of logic, such as an ASIC, which a test system cannot do.

Point 1 is basic and apparently obvious. Points (a) and (b) are particular details of this statement and (d), (e), (f), (h), (i), (j) and (k) are consequences of it. Point 2 is not quite so obvious. It draws attention to the fact that the simulator designer and implementor cannot anticipate all the mistakes that the user might make. Point 2 is thus a minimum statement. For many faults one would hope for a lot more. Points (c) and (g) are parts of 2.

Point 3 is a generalisation of (b). Points 5 to 9 are concerned with the user of the simulator. They draw attention to the dangers inherent in using a simulation. These dangers are basically three in number.

- The danger of checking against an incorrect or incomplete specification. It is essential to get the specification correct, and to check against that specification, only that specification and all of that specification.
- Implicit belief in the absolute correctness of the simulator. Every engineer should cultivate a suspicious mind.
- The temptation to adjust the simulator input data (design and/or test data) to obtain the required results with no 'complaints,' rather than making proper modifications to the design to eliminate real problems from the real hardware. The author has bitter experience of the latter approach.

Implementors of simulators also need to be aware of 'peripheral' matters as indicated by point 5. The fact is that input and output facilities and minimisation of the simulation cycle are *not* peripheral at all – they are critical to the sensible and efficient use of the simulator.

With a good modern design suite, points 1 and 3 can be assumed. No design system can give guarantees on point 2. It may well be that several different approaches are needed to get the greatest confidence. Thus a timing verification and a delay fault simulation may be needed as tools additional to the basic simulation. Problems also arise over the treatment of spikes – point 9, and again there is no 100% guarantee in any system.

Many design suites can now handle models at several levels of abstraction – point 4. VHDL demands it, and other languages give similar features. These languages enable models to be written at any level. Most design suites have model libraries for standard components. However, with increasing use of ASICs, some ability to write more complex models is required. Top down design is mandatory. A high level behavioural model and a lower level structural one can be written as separate VHDL ARCHITECTURES for the same ENTITY and their operation compared.

In the case of point 6, in relation to timing faults, the compromise preferred by this writer is for the system to press on with a 'best guess' of some sort. That will allow at least the possibility of the run giving more useful information. This presumes the inter-run period to be longer than the run time. If X values are available, the fault becomes very obvious, but they can propagate explosively. Where registers are involved, it is possible to do some resetting. Pressing on in the hope of a reset may be useful. There is a compromise to reach between trusting the user to be sensible and playing for safety – i.e. assuming the worst about the user.

10.2   Getting value from the simulator

10.2.1 Computer *aided* design

Systems provided to help in the design of engineering artefacts are known as computer **aided** design suites. There is a very great danger that the designer will become so immersed in getting the desired results from the simulator that the design achieved will be regarded as 'perfect.' The fact is that no design is perfect and that every manufacturing system yet devised will conspire to produce hardware which fails to perform properly, whether from deficiencies in design or in manufacturing.

The key to understanding the problem is the word *aided*. Computers will never be more than aids to the innate intelligence of the designer. It is when the designer begins to treat the tools as if they had intelligence that problems can arise. It is well from time to time to stand back and remind oneself of that fact, especially when under pressure to meet deadlines. These comments apply equally to so-called intelligent systems, since they also rely totally on fallible persons who designed them.

As with any tool, knowing its strengths and weaknesses means that the user can get a great deal more useful work from it. Steps can be taken to exploit the strengths and possibly work around the weaknesses. At least one can avoid being misled. Tools for electronic design can only realise their full potential if the user also understands the underlying electronics. In recent years much has been made of the potential of non-technical people being able to design chips. They can, but the best chips will be designed by people with the most thorough knowledge. This may seem obvious. It needs stating to counter over enthusiastic claims, sometimes from those who should know better.

Several of the procedures which enable good use to be made of a simulator are needed primarily for other purposes. These include the following.

- Structured design. This makes the design *understandable*, so that when mistakes occur they are more easily found. Because the design is understandable, it is easier to devise *tests* to check the functionality and testability. Because it is structured, it is possible to simulate each module separately. The advantage of this is that it is easier to identify problems in a relatively small module than in a relatively large system. Given some confidence in the correctness of the sub-units, simulating the system can concentrate on the module interaction alone.
- Testable design. The cost of testing is said to be 70% of the cost of chip design. Anything which will limit the cost of testing must be worthwhile, therefore. It will reduce both simulation time and the actual production testing costs. A testable design also means that it is relatively easy to fault simulate.

10.2.2 Models

The majority of simulators have a component model library covering many of the most popular integrated circuit families. In addition, most semiconductor vendors will provide models for their gates and macros. As a general rule, the details of how these work will not be available easily. A number of factors need to be considered.

Many simulators, quite properly, use an inertial delay model by default. This implies that any input signal shorter than the delay of the element (usually) will be removed. A great many irrelevant events are removed from the simulation and, more importantly, from the output which the user has to consider.

There are good theoretical reasons why the inertial delay model should be used, as was explained in Section 7.2. Whether the notion of pulse equal to the circuit delay is proper is much more questionable. To determine what should be used would require a circuit simulation of the basic circuit module. This is very unlikely to be available. As a result, this writer would expect simulations to be optimistic so far as spike removal is concerned, and, in the event of unexplained problems in hardware, this might provide an area to investigate. A simulation with all delays forced to transport delays will undoubtedly be pessimistic, but would give some clues to possible difficulties.

Writing one's own models for basic elements is not usually a viable option. Chapter 7 has demonstrated the difficulty of getting good models.

No claim is made for the completeness of the models described. Indeed, Chapter 7 was intended to point to the difficulties of writing and the limitations of the models rather than to encourage users to write them. Writing models is now an industry in itself. Apart from very specialist purposes or for high level, usually low accuracy models, the average user cannot hope to compete economically with the specialist companies. It is all the more important for the simulator user to be aware of possible model limitations.

Having said all that, there may be occasions when a model must be written. The modules of a system will be unique, and to model them before details are known will be the job of the designer. Accurate timing may or may not be required at various stages in the design. Where ASICs are designed, the user may well wish to write a more accurate model when detailed timing is known, possibly derived from a gate level simulation. If the models are well written and the software system is good, the detailed timing will be fed back automatically by back annotation. This will be especially necessary if a hardware modeller is not available. Even when it is, the timing checks will have to be written.

10.2.3 Testing functionality

Test generation is a matter of real concern. For device test, automatic generation is a possibility. However, the tests generated are not produced in a logical order suitable for determining the correctness of the design. Here is one area where a little thought can pay larger dividends.

Most systems will consist of a data path section or sections and means to control those section(s) to perform an appropriate one of several possible functions. The following is a minimum check.

- For each function of the system or sub-system, use several data sets. These should include the following.
  - Any initialisation.
  - Ensuring that every signal on each data path can change state.
  - Ensuring that adjacent or nearly adjacent bits are not stuck together. This refers to adjacency in relation to design errors, not manufacturing error. An example is a one character difference in a signal name.
  - Test on any special input data sets or a data set which gives exceptional results. Examples in floating point arithmetic might be a 'Not-a-Number' as an operand in the first case and an arithmetic overflow in the second.

- The user should be aware of the default initial values provided in the simulator. In a two-value system these could be all zeros (ones for flip-flop inverses). In this case, an initial reset to flip-flops will be a useless test. To test the master reset, the system must be put into an un-reset state by one or more prior input vectors chosen for the purpose. The signals can then be observed initialising. Such a situation is easily overlooked. A three-value system may not require this. The logic comes on with X values that are then initialised.
- The system should be checked with 'illegal' data input to ensure that it is not set into a dangerous state or one from which it cannot recover. Suppose that the system has an 'unused' function code. What happens if a user inadvertently supplies this? Will a memory be overwritten? Will an output be produced causing a railway signal to be set to green when it should be red?
- Skew. A variation of this can happen when input data changes. Unexpected or unusual conditions can arise. A good example is changing the address of a memory when the write signal is asserted. The address wires will all change at slightly different times in real life, but *they may well change at the same time in the simulator*. Under these circumstances, good data may be destroyed in the real hardware but probably not in the simulator. Some form of test to check for this and similar problems must be generated. The same effect happens with data on a register as a result of clock skew.

Suitable selection of data can lead to much better tests of the manufactured system. Few automatic test program generators attempt to find anything other than single s-a faults. A few may look for vector pairs to find stuck opens in MOS logic. Bridging faults are frequently ignored. Yet a lot can be done in this last case by selecting test data such that adjacent wires *in the final layout* have different values. These clearly are the most likely sources of bridging faults. Conversely, it is unlikely that wires at a distance will be shorted together if there is no local short. Finding such adjacencies is not a difficult problem for a program and should not take long to run. There are several references to the problems associated with bridging faults.

A severe problem with bridging faults is that they may introduce feedback where none existed before. This may lead to hyperactivity, which is often detected as such. If it does not, then the likelihood is that a new latch is produced and the normal tests will not find the fault. Herein lies the real difficulty with this type of fault.

10.3   Wires in high speed logic

Section 6.5.3 discussed the need for simulation of wires and how that might be achieved. There are two further complications where high speed logic is concerned.

Suppose the time for a signal to travel along a wire is greater than half the rise time (not rising delay here) of the signal. Suppose, further that the signal source resistance is a few ohms and the sink resistance is a few kilohms. Fig. 10.1 shows the waveforms at various points on the wire (Gosling 1985). If there is a gate attached to the wire at point B, the half-level signal may result in oscillation or other undesirable effects. Special rules are usually applied in the wiring program, or they must be checked afterwards. If the source and sink impedances are different from those suggested, and, in particular, if the load impedance is capacitive, the waveforms can become very complex and include spikes of significant length.

A further problem arises when two or more wires run close together for some distance. In this case both the electric (capacitive) and magnetic (inductive) coupling between the wires can result in there being a signal in one wire (the victim) as a result of a fast change in the other. This is known as **crosstalk**. The details are beyond the scope of this book except to note that they can be large enough to cause faulty operation.

Another difficulty occurs if the power line impedances are not sufficiently low, namely that the power/ground point at a gate can move to such an extent that there is an equivalent change of input state.

Again, a suitable layout program should be able to avoid pairs of wires

Fig. 10.1. Signals on a transmission line.

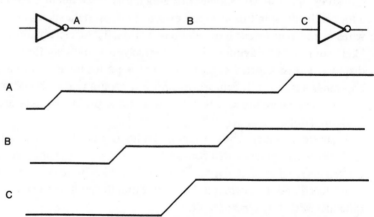

running close together for too long a distance. Detecting too high power line impedances is more difficult and may need a layout post processor.

10.4 Simulation accelerators

10.4.1 Point acceleration

A simulation accelerator is hardware specially designed to make a simulation run faster than it would run on a general purpose computer of equivalent technology. The accelerator will contain logic designed to do very rapidly operations which may take many instructions on the general purpose machine. As a rule, such hardware is of no use for any other purpose. Thus it is often called a **point accelerator** since it accelerates only one *point* in the total design process. In this sense, a hardware modeller is an accelerator. However, the term 'accelerator' is usually reserved for a more complete simulation system. A hardware modeller may well be incorporated into the accelerator.

Some writers have questioned the value of point accelerators. Consider Fig. 10.2. The height of the column represents the total design time from concept to finished product. The hatched portion represents the portion of the design time to be accelerated, in this case the simulation time. In Fig. 10.2(*a*) a 10-fold acceleration will reduce design time by 9%. Given the high price of special hardware, this is not cost-effective.

Consider, now, Fig. 10.2(*b*). A 10-fold speed up of the hatched portion will reduce design time by a factor of five. This may well be worth while.

Conventional wisdom suggests that simulation is more like Fig. 10.2(*b*)

Fig. 10.2. Value of a point accelerator.

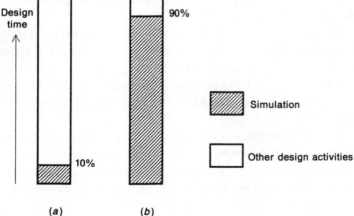

than (*a*). On the other hand, routing of PCBs is the other way about. Thus acceleration of simulation is worthwhile, whereas that of routing is not. Furthermore, the time to simulate can be reduced to such a degree that playing 'what if?' games becomes economically feasible. Faster simulation also allows more thorough testing of the functionality. As an example, a few years ago a company wished to simulate a new processor doing its bootstrap routine – several seconds worth of computing at several MIPS in real life. It took 20 days, 24 hours per day single user on a very powerful processor. Using a simulation accelerator with a speed-up of 1000 (which is possible), the same simulation takes half an hour.

10.4.2 Zycad engines

Several companies have developed accelerators, including Valid Logic (Realfast), Aida and Ikos. One of the best early (1980ish) machines was the Zycad Logic Evaluator (LE). It was capable of simulating 64K 'modelling elements' at a rate of 1M events per second per module, and could be equipped with 16 modules. Well-behaved designs are reported to have run even faster.

The machine operates in a manner similar to the event driven system of Fig. 6.7. There is no affected components list. Logic elements are limited to three-input one-output 'modelling elements.' Each input and the output has an associated delay. If two or three inputs change at the same time, the element has to be computed once for each input change. 31 different elements can be defined, many of them by the user. Higher level elements are simulated by means of hand crafted models using the modelling elements. A figure of about five to one saving over the 'real gates' can be achieved. The machine can simulate memory, but requires three real bits per simulated bit. Furthermore, this is taken from the memory available for input vectors and output reporting. Modelling uses three levels and four strengths, a 12-value arrangement.

This machine was excellent for the sort of long simulation mentioned above, especially as there were few input or output reports. It appeared that insertion of a primary input or recording of a signal caused a hiccup in the operation, as did a move from one time step to the next. Thus the engine worked best when there was a lot of activity at each time step and little input and output.

A further problem was the down load and compile times for networks. Figures published showed that these were generally less than alternative simulators, but were still in terms of minutes for a simulation of a few seconds. They were anything from about three to over 100 times as long as the simulation times.

In spite of these problems, it was undoubtedly a major step forward in simulation procedures. The machine was further developed to become the System Development Engine (SDE) in the mid-1980s. This was claimed to run at over 10^9 events per second. However, the number of modelling elements was still restricted to just over 1M. This was fine for chip design – its primary purpose – but not for system design. By 1991[1] a single board accelerator handling 256K elements at 2M events per second was available as a desk top machine, with larger models available for network resources. A desk top version has become available.

The LE was able to perform fault simulation by single fault insertion, if only because of its speed. However, it was soon realised that this was inadequate in relation to the size of the design checking problem. A development allowed the same machine to be used as a fault evaluator (FE) using the concurrent algorithm. One version of the fault evaluator was able to work with functional models, while another was able to perform at switch level.

10.4.3 Dazix Gigalogician

A primary problem with the Zycad style of accelerator is the limited modelling element (three inputs, one output) and the inability to handle storage in an efficient manner. Daisy Corp. – later Dazix – developed a machine which contained several processors (Kaul *et al.*, 1988). There is a hard wired processor capable of handling elements with up to five inputs including flip-flops and latches. There is a software processor to handle behavioural elements including memory, PLAs and devices specified in a modelling language – in this case Daisy's own, but it could now be VHDL. Finally, there is a physical modelling extension PMX, a hardware modeller. A *module* can hold up to 11 processors, of which up to five can be PMXs. A full machine may have 64 modules. A system such as this can change or upgrade processors since the simulation procedure and communication is independent of the processors.

It appears that each simulation processor holds data for specific elements and each module contains logic for a system capable of a full simulation. Thus there are two communications requirements.

- Data between elements.
- Data between modules.

This leads to a problem of allocation of elements to modules and

---

[1]   Electronic Product Design October 1991.

processors. Allocation strategies are always difficult and will almost always get some systems wrong.

The Dazix simulation system recognises the need to assist rapid resimulation during the design debug phase. All waveforms are monitored, therefore. Any desired waveform can be recalled during the debug phase of the work. The massive amounts of data of a large design are reduced partly by the use of high level models (fewer signals) and also by providing a means to send some of the results to disc. If the file allocation becomes full then the earliest results are deleted and replaced by more recent ones. Thus, if a fault is detected, the simulator can be arranged to stop itself and the events leading up to the fault are still recorded for the immediately preceding time periods. If the *event buffer* is large enough, the source of the fault will be detectable. In other words, the system operates in a manner similar to a logic analyser.

Full recompilation of a design is not necessary very often. When a fault is discovered, only the relevant module needs to be recompiled. Should the design modification require additional inputs or outputs to this module then the next higher level module must also be recompiled. If the module is used in several higher level sections of the design, they will each need to be recompiled. It will be necessary to recompile everything only very rarely.

The speed of simulating simple gates on the Gigalogician is probably less than that of the Zycad machines. However, it has better facilities for simulating higher level modules and hence systems, as opposed to single chips. It also has the very positive advantage of the event buffer giving faster fault finding. Recompilation should also be faster. These facilities necessarily lock the machine into the Dazix software and hardware suite. This is clearly a disadvantage for users who wish to use a range of tools or who have an investment in aids from other CAD vendors.

10.4.4 IBM machines

Earlier than the other machines, IBM built a simulator for their internal use. This used the compiled code algorithm rather than the event driven one. The description of Section 6.2 drew considerably on the detail of the Yorktown Simulation Engine (YSE). This was so powerful that it was said to be able to run programs in the simulation of the then latest 8-bit microprocessor faster than the hardware processor itself could run them. Of course, it was much larger and more expensive! IBM have since developed several other engines from this (Beece *et al.* 1988).

10.4.5 Assessment

The primary advantages of hardware simulators are the following.

- There is no operating system as there would be in a general purpose computer.
- The hardware is tuned to the algorithm and hence must be faster.
- The machine will use pipelines to overlap operations and parallel processors to effect higher processing speed.
- There is a large real memory and no virtual memory. As data is required in an unordered fashion at all times, virtual memory techniques will lead to thrashing and loss of speed.

The primary problem with hardware simulators is that they have a fixed size and hence there is always some network that will not fit into it. With machines based on relatively simple logical elements, there is no way round that, though some sort of 'high level' model is possible. To get round the problem, it is necessary for the simulator to be able to handle behavioural and hardware models. Under these circumstances, if the network becomes too large then part of it can be modelled at a higher level, thus enabling a larger network to be simulated.

Hardware simulators are only of real use for large designs and long simulations, preferably with little primary input or output. Blank (1984) points out that, on small designs, the down loading and other overheads of a hardware engine may lead to it being up to 10 times *slower* than a software simulator. He quotes a 500K gate network with a 100 instruction sequence taking 49 min on an accelerator but only 4.5 min on a software simulator. When the number of instructions was raised to 10^6, however, the simulation took 66 min on an accelerator and 250 h on the software simulator.

10.5   Whither now?

It is clear that, with ever expanding design sizes, some means of checking a design against specification is essential. Furthermore, it is clear that a mixed level simulation is also vital to enable system designs to be checked while some modules are only partly specified, as well as to reduce the problem size when checking some modules at the most detailed levels. A run time of 20 days for a processor chip has already been mentioned. In 1985 Zycad quoted the run time of a conventional simulator on a small design (1700 gates, 8600 test vectors) as half an hour. The requirements of the simulator increase according to the gate count to some power greater than one – often two or even three. This is faster than the rate of increase of conventional computing power. Fault simulation for test grading is even worse.

Clearly simulation accelerators are necessary for the largest designs. On the 1700 gate design above Zycad quoted the Logic Evaluator with a compile time of 42 s and a run time of 7 s. A design of 8000 elements had a compile time of nearly 4 min and a run time of under two *seconds*. The speed improvements due to acceleration are impressive. However, it will be appreciated that the Dazix emphasis on help with debugging and on incremental compilation is very necessary.

Other hierarchical methods of reducing the complexity of both simulation and fault simulation have been mentioned. These must be developed further.

Perhaps the main problem for which there is as yet no real solution is that of how to prepare a comprehensive specification for a system to be designed and then to ensure that the design meets that specification. Automatic test program generation, as it is known in 1992, is aimed at testing manufactured samples of a known good design. It is no use for determining whether the design meets the specification. This is still very much up to the designer's ingenuity and honesty.

One suggested approach to the problem is in the area of formal specification and theorem proving. Formal specification has been used successfully on a number of occasions. Some examples of design proving with these ideas appear to be little more than alternative simulation methods. They do little to check timing and can handle relatively small parts of the logic. It is hoped that there might be something better on the way.

Another approach is the use of automatic design from specification. The specification is verified formally. The behaviour should then be **correct by construction**. However, the designs will almost certainly be less compact than if expert human intelligence were used. If the difference is only 10 or 20% and other constraints are met, the improved reliability of the design is well worth the cost. Behavioural simulation of such designs is unnecessary. Whether timing verification is necessary would be dependent on the design system. If timing were tight, some manual intervention might be necessary, but that would negate the advantages of the automatic design, and simulation would be essential.

Even with automatic design, test program generation and fault simulation are still required. Both these tasks are quite large and, in their present form, limited in what they can do (single stuck at faults, for example). The only solution would be to use some form of parallelism. It might be possible to partition the possible faults among several processors and then to generate tests and fault simulate on each processor independently. Some faults might get tested several times but every fault would be checked once. Even if all possible faults could be detected by a test set, it is likely that the set would be so large that testing, which cannot use parallelism, would be uneconomic. Thus there is a compromise to be struck between cost and fault cover. Full fault cover can be reserved for devices intended for safety critical applications and would carry a premium on the price.

Finally, which simulator? This author cannot answer that. It depends a great deal on the user's philosophy and approach. The important thing is to be conscious of the problems. Greer (1987) gives a list of matters to be considered when choosing a simulator. These include the following.

- Operating capabilities; levels and strengths; interactive break on condition and forced values; incremental compilation; save and restart.
- Timing: unit, zero, variable delays; random delay assignment; rising and falling delays; inertial delays and spike filtering; ambiguity models; load dependent delays; set-up and hold checks.
- Hardware constructs: transmission gates; tristate; pull-up/down resistors; wired logic; ECL with true/complement delay; switch level charge storage.
- Logic evaluation: real and potential spike and hazard reports; unit delay oscillation; hazard with reconvergent fan-out; tristate bus connections.
- Fault simulation: s-a-$1/0$; bridging faults; wire open faults; fault reporting; probabilistic analysis; delay faults.

Not every user wants all these facilities. Each person must select that set of facilities that are important in his/her own circumstances.

Appendix

The algorithm of Prog. 7.1 was written for a hardware simulator which makes no assumptions. If it is run as shown on a commercial software simulator, it may not work without modification. There are two reasons.

The first is that Prog. 7.1 assumes that all input changes at a given simulation time are evaluated together – that is, an affected components list is used. If that is not the case and the s change is evaluated before an input change at the same time then incorrect results are obtained.

To avoid this advantage can be taken of the VHDL concept of deltas. The fragment of Prog. 7.1 is encased in a further CASE statement of the form

```
CASE s IS
    WHEN '1' = >
        s < = TRANSPORT '0';
                -- happens after one delta as s is a SIGNAL
    WHEN '0' = >
        Prog. 7.1
END CASE;
```

If the input change comes from the event memory before the s change, then s is '0' and the effect of the change is evaluated. A following s change will then fail on the time delay test. On the other hand, if the s event appears first, all that happens is that s is predicted to change to '0' after a delta. The input change then occurs, but there is no evaluation as s is '1.' A second s < = '0' prediction is made, but this deletes the previous one (see below).

Next a delta time step is made and s < = '0' is found. The logic is evaluated with the input now at the required value.

The second problem arises because most simulators have a built in 'buffer' which understands the concept of TRANSPORT and INERTIAL delays. In particular, the prediction of $s < =$ TRANSPORT '0' will remove any other predictions for s made *before* the current time but due to be realised *after* current time as shown in Fig. A1. Hence the later s has to be replaced. A statement can be added to the "WHEN '1' = >" part of the new case clause above as shown below.

$$\text{IF } ((NOW - T) < (t_r - t_f)) \text{ THEN}$$
$$s < = \text{TRANSPORT '1' AFTER } (t_r - t_f - (NOW - T));$$
$$\text{END IF};$$

If there are several s predictions ahead, then all will be removed. Only the last will be replaced. It can be shown that, for this 'simple' buffer, that is adequate, but in some more complex logic block it may not be.

These mechanisms have been used to test the state diagrams of Figs. 7.7, 7.13 and 7.14 on a commercial simulator and have been found to work.

Fig. A1. Illustration of effects of transport delay.

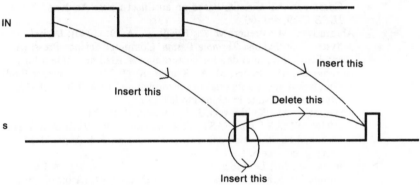

References

Abadir, M. S.; Breuer, M. A. (1985): 'A Knowledege-Based System for Designing Testable VLSI Chips.' *IEEE Design and Test of Computers*, **2–4**, Aug., 56–68.

Abramovici, M.; Breuer, M.A. (1980): 'Multiple Fault Diagnosis on Combinational Circuits Based on an Effect-Cause Analysis.' *Trans IEEE* **C-29**, 451–60.

Abramovici, M.; Breuer, M. A.; Friedman, A. D. (1990): *Digital Systems Testing and Testable Design.* Computer Science Press. (An excellent book, intended for postgraduates. Assumes quite a lot.)

Abramovici, M.; Breuer, M. A.; Kumar, K. (1977): 'Concurrent Fault Simulation and Functional Level Modelling.' *14th DAC*, 128–37. (Rudimentary state machine modelling.)

Abramovici, M.; Kulikowski, J. J.; Menon, P. R.; Miller, D. T. (1986): 'SMART and FAST; Test Generation for VLSI Scan-Design Circuits.' *IEEE Design and Test of Computers*, **3–4**, Aug., 43–54. (Very good – covers a lot of ideas.)

Abramovici, M.; Levendel, Y. H.: Menon, P. R. (1983a): 'A Logic Simulation Machine.' *Trans IEEE* **CAD-2**, 82–94. (A bit old but some useful ideas.)

Abramovici, M.; Menon, P. R.; Miller, D. T. (1984): 'Critical Path Tracing – an Alternative to Fault Simulation.' *IEEE Design and Test of Computers*, **1–1**, Feb., 83–92.

Abramovici, M.; Menon, P. R. (1985): 'A Practical Approach to Fault Simulation and Test Generation for Bridging Faults.' *Trans IEEE* **C-34**, 658–63.

Abramovici, M.; Rajan, K. B.; Miller, D. T. (1992): 'FREEZE!: A New Approach for Testing Sequential Circuits.' *29th DAC*, 22–5. (Finds multiple tests for each sequential state. Very useful for circuits not fully scan designed.)

Agarwal, V. K.; Fung, A. S. F. (1981): 'Multiple Fault Testing of Large Circuits by Single Fault Test Sets.' *Trans IEEE* **C-30**, 855–65. (Mainly for fan-out free circuits, but interesting. Circuits not large by 1990 standards.)

Agrawal, P.; Dally, W. J. (1990): 'A Hardware Logic Simulation System.' *Trans IEEE* **CAD-9**, 19–29. Extensions to MARS (see below).

Agrawal, P.; Dally, W. J.; Ezzat, A. K.; Fischer, W.C.; Jagadish, H. V.; Krishnakumar, A. S. (1987): 'Architecture and Design of the MARS Hardware Accelerator.' *24th DAC*, 101–7. Also *IEEE Design and Test of Computers*, 1987 **4–5**, Oct. 28–36.

Agrawal, P.; Dally, W. J.; Tutundjian, R. (1989): 'Algorithms for Accuracy Enhancement in a Hardware Logic Simulator.' *26th DAC*, 645–8.

Antreich, K. J.; Schulz, M. H. (1987): 'Accelerated Fault Simulation and Fault Grading in Combinational Circuits.' *Trans IEEE* **CAD-6**, 704–12.

Armstrong, D. B. (1972): 'A Deductive Method for Simulating Faults in Logic Circuits.' *Trans IEEE* **C-21**, 464–71.

Armstrong, J. R. (1984): 'Chip-Level Modeling of LSI Devices.' *Trans IEEE* **CAD-3**, 288–97. (Good ideas. Also fault modelling.)

Barrett, G. (1987): 'Formal Methods Applied to Floating Point Number Systems.' *Tech Mono. PRG-58*. University of Oxford.

Baschiera, D.; Coutois, B. (1984): 'Testing CMOS – a Challenge.' *VLSI Design*, **5**, Oct., 58–62.

Becker, B.; Hahn, R.; Kreiger, R.; Sparmann, U. (1991): 'Structure Based Methods for Parallel Pattern Fault Simulation in Combinational Circuits.' *EDAC 91*, 497–502. (Speed-up of PPSFP.)

Beece, D. K.; Deibert, G.; Papp, G.; Villante, F. (1988): 'The IBM Engineering Verification Engine (EVE).' *25th DAC*, 218–24.

Benkoski, J.; Stewart, R. B. (1991): 'TATTOO: An Industrial Timing Analyser with False Path Elimination and Test Pattern Generation.' *EDAC 91*, 256–60. (From SGS-Thompson.)

Benkoski, J.; Strojwas, A. J. (1987): 'A New Approach to Hierarchical and Statistical Timing Simulations.' *Trans IEEE* **CAD-6**, 1039–52.

Bennetts, R. G. (1984): 'Design of Testable Logic Circuits.' Addison-Wesley.

Blank, T. (1984): 'A Survey of Hardware Accelerators used in Computer Aided Design.' *IEEE Design and Test of Computers*, **1–3**, Aug., 21–39. (A bit old now but covers some good things.)

Bolender, E.; Lipp, H. M. (1992): 'Timing Verification: A New Understanding of False Paths.' *EDAC 92*, 383–7. (Circuit replicated for every fan-out; all delays transferred to the circuit inputs. Not applicable to asymetric delays.)

Brayton, R. K.; Hachtel, G. D.; McMullen, C. T.; Sangiovanni-Vincentelli, A. L. (1984): 'Logic Minimization Algorithms for VLSI Synthesis.' Kluwer Academic Publishers.

Breuer, M. A; Friedman, A. D. (1980): 'Fundamental Level Primitives in Test Generation.' *Trans IEEE* **C-29**, 223–35. (Very useful. Use of vector sequences of input, output and internal states for high level primitives. Examples of D-drive and justification for 74194 shift register and 74190 counter.)

Bryant, R. E. (1984): 'A Switch-Level Model and Simulator for MOS Digital Systems.' *Trans IEEE* **C-33**, 160–77.

Bryant, R. E.; Beatty, D.; Brace, K.; Cho, K.; Scheffler, T. (1987): 'COSMOS: A Compiled Simulator for MOS Circuits.' *24th DAC*, 9–16.

Camurati, P.; Prinetto, P. (1988): 'Formal Verification of Hardware Correctness: Introduction and Survey of Current Research.' Computer **21-7**, 8–19.

Cha, C. W.; Donath, W. E.; Özgüner, F. (1978): '9-V (valued) Algorithm for Test Pattern Generation of Combinational Digital Circuits.' *Trans IEEE* **C-27**, 193–200. (Modification to Muth (1976). Slow on small circuits. Clear example. Limited to primitive gates.)

Chakravarty, S. (1992): 'Heuristics for Computing Robust Tests for Stuck-Open Faults from Stuck-At Tests.' *EDAC 92*, 416–20. ('Robust' means not affected by circuit delays. Fault collapsing cannot be used – may lose robustness. Gates classified so can easily get the required stuck open tests in over 70% of cases.)

Chamberlain, R. D.; Franklin, M. A. (1986): 'Collecting Data About Logic Simulation.' *Trans IEEE* **CAD-5**, 405–12.

Chandra, S. J.; Patel, J. H. (1987): 'A Hierarchical Approach to Test Generation.' *24th DAC*, 495–501.

Chandra, S. J.; Patel, J. H. (1989): 'Experimental Evaluation of Testability Measures for Test Generation.' *Trans IEEE* **CAD-8**, 93–7. (Comparison of several testability measures on sample circuits.)

Chappell, S. G.; Elmendorf, C.H.; Schmidt, L. D. (1974): "LAMP": Logic Circuit Simulators.' *BSTJ*, **53**, 1451–76. (Review of *six* simulators in the LAMP system at Bell Labs. Interesting figures. Other papers on LAMP in the same issue are of interest.)

Cheng, W-T. (1988): 'SPLIT Circuit Model for Test Generation.' *25th DAC*, 96–101. (Review of five- and nine-value algorithms.)

Cheng, W-T.; Yu, M-L. (1989): 'Differential Fault Simulation – a Fast Method Using Minimal Memory.' *26th DAC*, 424–8. (Variation on concurrent.)

Cherry, J. J. (1988): 'Pearl: A CMOS Timing Analyser.' *25th DAC*, 148–53.

Choi, K.; Hwang, S. Y.; Blank, T. (1988): 'Incremental-in-Time Algorithm for Digital Simulation.' *25th DAC*, 501–5. (An alternative approach to simulation.)

Cirit, M. A. (1988): 'Switch Level Random Pattern Testability Analysis.' *25th DAC*, 587–90. (Another assessment of CY and OY with switch level in mind. Could probably be extended to gate level.)

Coelho, D. R. (1984): 'Behavioural Simulation of LSI and VLSI Circuits.' *VLSI Des*, **5**, 42–51. (Refers to hierarchical and behavioural simulation and *why*.)

Cohn, A.: 'A Proof of Correctness of the VIPER Microprocessor – the First Level.' Report of University of Cambridge Computer Lab.

d'Abreau, M. A.; Thompson, E. W. (1980): 'An Accurate Functional Level Concurrent Fault Simulator.' *17th DAC*, 210–17.

Demba, S.; Ulrich, E.; Lentz, K. P.; Giramma, D. (1990): 'Experiences with Concurrent Fault Simulation of Diagnostic Programs.' *Trans IEEE* **CAD-9**, 621–8.

Denneau, M. M. (1982): 'The Yorktown Simulation Engine.' *19th DAC*, 55–9. (See also Pfister (1982) and Kronstadt (1982).)

Denneau, M. M.; Kronstadt, E.; Pfister, G. (1983): 'Design and Implementation of a Software Simulation Engine.' *Computer Aided Design*, **15** 123–30.

Dervisoglu, B. I. (1990): 'Application of Scan Hardware and Software for Debug Diagnostics in a Workstation Environment.' *Trans IEEE* **CAD-9**, 612–20. (Scan design of a real machine – the Apollo

DN10000. Eight ports per board, pseudo-random pattern generator and signature compactors. ≤10% overhead.)

Deutsch, S.; Lhotak, E. (1990): 'Concurrent Hardware Modelling Speeds Simulation.' *High Performance Systems*, June, 41–5. (Racal-Redac hardware modeller – also for fault simulation.)

Devadas, S.; Keutzer, K. (1990): 'Synthesis and Optimisation Procedures for Robustly Delay-Fault Testable Combinational Logic Circuits.' *27th DAC*, 221–7. (Two-level circuits extended by hierarchy. Useful ideas.)

Du, D. H. C.; Yen, S. H. C.; Ghanta, S. (1989): 'On the General False Path Problem in Timing Analysis.' *26th DAC*, 555–60. (Examples of static analysis giving incorrect false paths.)

Elziq, Y. M. (1981): 'Automatic Test Generation for Stuck-Open Faults in CMOS VLSI.' *18th DAC*, 347–54.

Evanczuk, S. (1989): 'IEEE 1149.1; A Designer's Reference.' *High Performance Systems*, Aug., 52–60. (Description of IEEE 1149.1 Boundary Scan standard.)

Fazakerly, W. B.; Smith, R. P. (1988): 'Advanced Modeling Techniques for Logic Simulation.' *VLSI Systems Design, Semicustom Design Guide*.

Freeman, S. (1988): 'Test Generation for Data-Path Logic – the *F*-path Method.' *IEEE J SSC*, **23**, 421–7.

Fuchs, K.; Fink, F.; Schulz, M. H. (1991): 'DYNAMITE: An Efficient Automatic Test Pattern Generation System for Path Delay Faults.' *Trans IEEE* **CAD-10**, 1323–35. (Good explanation of robust/non-robust tests.)

Fujiwara, H.; Shimono, T. (1983): 'On the Acceleration of Test Generation Algorithms.' *Trans IEEE* **C-32**, 1137–44. (Description of FAN.)

Gai, S.; Montessoro, P. L. (1991): 'The Fault Dropping Problem in Concurrent Event-Driven Simulation.' *Trans IEEE* **CAD-10**, 968–71. (Recommends fault dropping every few tests and shows that it is cost effective.)

Gai, S.; Montessoro, P. L.; Somenzi, F. (1988): 'MOZART: A Concurrent Multilevel Simulator.' *Trans IEEE* **CAD-7**, 1005–16.

Gai, S.; Somenzi, F.; Ulrich, E. (*1987*): 'Advances in Concurrent Multilevel Simulation.' *Trans IEEE* **CAD-6**, 1006–12. (In MOZART.)

Gaiotti, S.; Dagenais, M. R.; Rumin, N. C. (1989): 'Worst-case Delay Estimation of Transistor Groups.' *26th DAC*, 491–6.

Gheewala, T. (1989); 'Cross check: A Cell Based VLSI Testability Solution.' *26th DAC*, 706–9. (Small blocks of logic with many test points connected through a small pass transistor. Thus uses only 5% of chip area and loading effect is small. Quotes 26 faults for two-input CMOS gate and 150 for a flip-flop.)

Ghosh, A.; Devadas, S.; Newton, A. R. (1991): 'Test Generation and Verification for Highly Sequential Circuits.' *Trans. IEEE* **CAD-10**, 652–67.

Gielen, G.; Liu, E.; Sangiovanni-Vincentelli, A.; Gray, P. (1992): 'Analog Behavioural Models for Simulation and Synthesis of Mixed-Signal Systems.' *EDAC 92*, 464–8.

Goel, P. (1981): 'An Implicit Enumeration Algorithm to Generate Tests for Combinational Logic Circuits.' *Trans IEEE* **C-30**, 215–22.

(PODEM original.)

Goel, P.; Lichaa, H.; Rosser, T. E.; Stroh, T. J.; Eichelberger, T. J. (1980): 'LSSD Fault Simulation Using Conjunctive Combinational and Sequential Methods.' *Proc IEEE Test Symposium*, 371–6.

Goldstein, L. H. (1979): 'Controllability/Observability Analysis of Digital Circuits.' *Trans IEEE* **CAS-26**, 685–93. (SCOAP again, but comparison of two methods of calculating CY and OY.)

Goldstein, L. H.; Thigpen, E. L. (1980): 'SCOAP: Sandia Controllability/Observability Analysis Program.' *17th DAC*, 190–6.

Gosling, J. B. (1985): 'Digital Timing Circuits.' Edward Arnold (now Hodder).

Greer, D. L. (1987): 'The Quick Simulator Benchmark.' *VLSI Design*, **8**, Nov., 40–57.

Hajjar, A.; Marbot, R.; Greiner, A.; Kiani, P. (1991): 'TAS: An Accurate Timing Analyser for CMOS VLSI.' *EDAC 91*, 261–5.

Harding, W. (1989): 'System Simulation Assures that Chips Play Together.' *Computer Design*, Aug. 1st, 70–84. (Excellent article covering need for high level and system simulation as well as gate level.)

Hayes, J. P. (1986): 'Digital Simulation with Multiple Logic Values.' *Trans IEEE* **CAD-5**, 274–83.

Hitchcock, R. B. (1982): 'Timing Verification and the Timing Analysis Program.' *19th DAC*, 594–604.

Hitchcock, R. B.; Smith, G. L.; Cheng, D. D. (1982): 'Timing Analysis of Computer Hardware.' *IBM J. Res and Devt*, **26**, 100–05. (Finds all problems *except those due to engineers!* Good paper.)

HITIME Reference Manual, GENRAD, Fareham, Hants, UK.

Hodge, S. (1990): 'Simultaneous Engineering and EDA Tools.' *Electronic Product Design*, Feb., 33–6. (Several very vital statements.)

Hollingworth, P. (1991): 'The rise of VHDL: 1076 and All That.' *IEE Review*, **37**, 139–42. (Good introduction at a low technical level.)

Hughes, J. L. A. (1988): 'Multiple Fault Detection Using Single Fault Test Sets.' *Trans IEEE* **CAD-7**, 100–08.

Illman, R.; Clarke, S. (1990): 'Built-in Self Test of the Macrolan Chip.' *IEEE Design and Test of Computers*, **7-2**, Apr., 29–40. (A real application of BIST, with some useful data.)

Ishiura, N.; Takahashi, M.; Yajima, S. (1989): 'Time-Symbolic Simulation for Accurate Timing Verification of Asynchronous Behaviour of Logic Circuits.' *26th DAC*, 497–502.

Jain, S. K.; Agrawal, V. D. (1984): 'STAFAN: An Alternative to Fault Simulation.' *21st DAC*, 18–23.

Jouppi, N. P. (1987): 'Timing Analysis and Performance Improvement of MOS VLSI Designs.' *Trans IEEE* **CAD-6**, 650–65.

Ju, Y-C.; Saleh, R. A. (1991): 'Incremental Techniques for Identification of Statically Sensitizable Critical Paths.' *28th DAC*, 541–6. (Getting rid of false paths; paths in order of non-increasing delay.)

Kaul, S.; Jacobson, N.; Livnat, I. (1988): 'Multiprocessor Accelerator for Mixed-Level Simulation.' *VLSI System Design*, **9**, May, 64–71. (Excellent paper giving much good philosophy on simulation as well as a description of the Gigalogician.)

Keller, B. L.; Carlson, D. P.; Maloney, W. B. (1991): 'The Compiled

Logic Simulator.' *IEEE Design and Test of Computers*, **8-1**, Mar., 21–34.

Kelly, N. F.; Heideman, W. P.; Papamarcos, M. S. (1989): 'Hardware Modeler Spans Multiple Environments.' *High Performance Systems*, April, 24–40.

Levendel, Y. L.; Menon, P. R.; Miller, C. E. (1981): 'Accurate Logic Simulation Models for TTL Totempole and MOS Gates and Tristate Devices.' *BSTJ*, **60**, 1271–87.

Kirkland, T.; Flores, V. (1983): 'Software Checks Testability and Generates Tests of VLSI Design.' *International Electronics*, **56**, 210–24. (Calculates CY/OY/TY on an actual circuit; demonstrates how to improve testability. Simply put and helpful.)

Kirkland, T.; Mercer, M. R. (1988): 'Algorithms for Automatic Test Pattern Generation.' *IEEE Design and Test of Computers*, **5-3**, June, 43–55. (Review of D-algorithm, PODEM and FAN·)

Knapp, D. W.; Winslett, M. (1992): 'A Prescriptive Formal Model for Data-Path Hardware.' *Trans IEEE* **CAD-11**, 158–84. (Formal methods; Claims to point to repairs to a design that does not meet the specification.)

Kodandapani, K. L; Pradhan, D. K. (1980): 'Undetectability of Bridging Faults and Validity of Stuck-at Fault Test Sets.' *Trans IEEE* **C-29**, 55–9.

Koike, N.; Ohmori, K.; Sasaki, T. (1985): 'HAL: A High Speed Logic Simulation Machine.' *IEEE Design and Test of Computers*, **2-5**, Oct., 61–73.

Kronstadt, E.; Pfister, G. (1982): 'Software Support for the Yorktown Simulation Engine.' *19th DAC*, 60–4. (See also Pfister (1982) and Denneau (1982).)

Landman, B. S.; Russo, R. I. (1971): 'On a Pin Versus Block Relationship for Partitions of Logic Graphs.' *Trans IEEE* **C-20**, 1469–79.

Lee, H. K.; Ha, D. S.; Kim, K. (1989): 'Test Pattern Generation for Stuck-Open Faults Using Stuck-at Test Sets in CMOS Combinational Circuits.' *26th DAC*, 345–50.

Lesser, J. D.; Shedletsky, J. J. (1980): 'An Experimental Delay Test Generator for LSI Logic.' *Trans IEEE* **C-29**, 235–48.

Levendel, Y. H.; Menon, P. R. (1982): 'Test Generation Algorithms for Computer Hardware Description Languages.' *Trans IEEE* **C-31**, 577–88. (Propagation D-cubes for several high level functions and attempts to formalise.)

Levendel, Y. H.; Menon, P. R. (1981): 'Fault Simulation Methods – Extensions and Comparison.' *BSTJ*, **60**, 2235–58. (Useful review of methods of fault simulation.)

Lewis, D. M. (1991): 'A Hierarchical Compiled Code Event-Driven Logic Simulator.' *Trans IEEE*, **CAD-10**, 726–37. (Some interest, but in the end unit delay nullifies the strength of the event driven approach.)

Lewis, D. M. (1992): 'A Compiled-Code Hardware Accelerator for Circuit Simulation.' *Trans IEEE* **CAD-11**, 555–65. (AWESIM III.)

Li, W. N.; Reddy, S. M.; Sahni, S. (1988). 'On Path Selection in Combinational Logic Circuits.' *25th DAC*, 142–7. (Paths selected for delay testing. For any edge in graph, select longest path including that edge.)

Lioy, A. (1992): 'Advanced Fault Collapsing.' *IEEE Design and Test of Computers*, **9-1**, Mar., 64–71. (Derives a formal method of fault collapsing to be used before any test generation.)

Lo, C-Y.; Nham, H. N.; Bose, A. K. (1987): 'Algorithms for an Advanced Fault Simulation System in MOTIS.' *Trans IEEE* **CAD-6**, 232–40. (Concurrent simulation applied to switch level. Good example.)

Maly, W. (1987): 'Realistic Fault Modelling for VLSI Testing.' *24th DAC*, 173–80. (Relates to sources of faults in silicon.)

Markowitz, R.; Wild, A. (1992): 'Accurate Delay Models for ECL Logic Synthesis.' *EDAC 92*, 97–101. (Notes that even metal wires at 1 μm are resistive. 'Synthesis' in the title seems irrelevant.)

Maurer, P.M. (1991): 'Scheduling Blocks for Hierarchical Complied Simulation of Combinational Circuits.' *Trans IEEE*, **CAD-10**, 184–92. (Hierarchy reduces size of code in LCC. Balance against longer simulation time due to more complex evaluation, i.e. uses flattened models for blocks, not high level models.)

McCluskey, E. J. (1985): 'Built-in Self-Test Techniques.' *IEEE Design and Test of Computers*, **2-2**, April, 21–8. (Useful overview; lot of references.)

McGeer, P. C.; Brayton, R. K. (1989): 'Efficient Algorithms for Computing the Longest Viable Path in a Combinational Network.' *26th DAC*, 561–7.

McWilliams, T. M. (1980): 'Verification of Timing Constraints on Large Digital Systems,' *17th DAC*, 139–47.

Moorby, P. R. (1983): 'Fault Simulation Using Parallel Value Lists.' *ICCAD 83*, 101–2.

Murray, B. T.; Hayes, J. P. (1990): 'Hierarchical Test Generation Using Precomputed Tests for Modules.' *Trans IEEE* **CAD-9**, 594–603. (Calculates CY and OY for modules of a more complex type than in this text.)

Muth, P. (1976): 'A Nine-Valued Circuit model for Test generation.' *Trans IEEE* **C-25**, 630–6.

Najm, F. N.; Hajj, I. N. (1990): 'The Complexity of Fault Detection in MOS VLSI Circuits.' *Trans IEEE* **CAD-9**, 995–1001. (Very comprehensive cover of MOS faults.)

Niermann, T.; Patel, J.H. (1991): 'HITEC: A Test Generation Package for Sequential Circuits.' *EDAC 91*, 214–18.

Nishida, T.; Miyamoto, S.; Kozawa, T.; Satoh, K. (1987): 'RFSIM: Reduced fault Simulator.' *Trans IEEE* **CAD-6**, 392–402.

Noujain, S. E. (1984): 'A Structural Approach to Test Vector Generation.' *ICCD*, 757.

Ousterhout, J. K. (1985): 'A Switch-Level Timing Verifier for Digital MOS VLSI.' *Trans IEEE* **CAD-4**, 336–49. (CRYSTAL.)

Perremans, S.; Claesen, L.; de Man, H. (1989): 'Static Timing Analysis of Dynamically Sensitizable Paths.' *26th DAC*, 568–73.

Perry, D. L. (1991): 'VHDL' McGraw Hill.

Pfister, G. F. (1982): 'The Yorktown Simulation Engine: Introduction.' *19th DAC*, 51–4. (See also Denneau (1982) and Kronstadt (1982).)

Putzolu, G. P.; Roth, J. P. (1971): 'A Heuristic Algorithm for the Testing of Asynchronous Circuits.' *Trans IEEE* **C-20**, 639–47. (Good paper, but doesn't guarantee test if one exists.)

Rajsuman, R.; Jayasumana, A. P.; Malaiya, Y. K. (1989): 'CMOS Stuck-Open Fault Detection Using Single Test Patterns.' *26th DAC*, 714–17. (Notes effect of glitches on two-test patterns for stuck open faults.)

Reddy, M. K.; Reddy, S. M. (1986): 'Detecting FET Stuck-Open Faults in CMOS Latches and Flip-flops.' *IEEE Design and Test of Computers*, **3-5**, Oct., 17–26. (How to design testable flip-flops and latches.)

Reddy, S. M.; Agrawal, V. D.; Jain, S. K. (1984): 'A Gate Level Model for CMOS Combinational Logic Circuits with Application to Fault Detection.' *21st DAC*, 504–9. (Models n and p FETs separately and handles stuck-open, stuck closed.)

Rezac, R. R.; Smith, L. T. (1984): 'Methodology for and Results from the use of a Hardware Logic Simulation Engine.' *ICCD*, 457–61. (Useful early view on use of Zycad. See Smith (1986).)

Rogers, W. A.; Guzolek, J. F.; Abraham, J. A. (1987): 'Concurrent Hierarchical Fault Simulation: A Performance Model and Two Optimizations.' *Trans IEEE* **CAD-6**, 848–62.

Roth, J. P. (1966): 'Diagnosis of Automata Failures: A Calculus and a Method.' *IBM J Res and Devt* **10**, 278–91.

Rumsey, M.; Sackett, J. (1989): 'An ASIC Methodology for Mixed Analogue-Digital Simulation.' *26th DAC*, 618–21. (Behavioural models in analogue, including FFTs.)

Runyan, S. (1989): 'Mixed Signal Design: Mounting the Barrier.' *High Performance Systems*, May, 26–43. (Useful view of Montage – genuine mixed level simulator.)

Russell, G.; Sayers, I. (1989): *Advanced Simulation and Test Methodologies for VLSI Design*. van Nostrand Reinhold. (Good book, especially on design for test, but not hot on simulation.)

Sakallah, K. A.; Director, S. W. (1985): 'Samson2: An Event Driven VLSI Circuit Simulator.' *Trans IEEE* **CAD-4**, 668–84.

Sarfert, T. M.; Markgraf, R. G.; Schulz, M. H.; Trischler, E. (1992): 'A Hierarchical Test Pattern Generation System Based on High-Level Primitives.' *Trans IEEE* **CAD-11**, 34–44.

Savir, J. (1983): 'Good Controllability and Observability do not Guarantee Good Testability.' *Trans IEEE* **C-32**, 1198–200.

Schultz, M. H.; Trischler, E.; Sarfert, T. H. (1988): 'SOCRATES: A Highly Efficient Automatic Test Pattern Generator.' *Trans IEEE* **CAD-7**, 126–37.

Schulz, M. H.; Fink, M. H.; Fuchs, K. (1989): 'Parallel Pattern Fault Simulation of Path Delay Faults.' *26th DAC*, 357–63.

Shen, J. P.; Hirschhorn, D. (1987): 'Switch Level Techniques.' *IEEE Design and Test of Computers*, **4-4**, Aug., 15–16 (Guest Editorial).

Smith, L. T.; Rezac, R. R. (1984): 'Methodology for and Results from the use of a Hardware Logic Simulation Engine for Fault Simulation.' *Int Test Conf*, 224–8. (See Rezac and Smith (1984).)

Smith, R. J. (1986): 'Fundamentals of Parallel Logic Simulation.' *23rd DAC*, 2–12. (Multiple processor simulation. Useful paper.)

Smith, S. P.; Mercer, M. R.; Brock, B. (1987): 'Demand Driven Simulation: BACKSIM.' *24th DAC*, 181–7. (An alternative simulation method.)

Somenzi, F.; Gai, S.; Mezzalama, M.; Prinetto, P. (1985): 'Testing Strategy and Techniques for Macro-Based Circuits.' *Trans IEEE*

C-34, 85–90. (Useful. D-algorithm applied to PLAs. Use of CY and OY.)

Son, K. (1985): 'Fault Simulation with the Parallel Value List Algorithm.' *VLSI Syst. Des*, **6** Dec., 36–43. (HILO algorithm.)

Soule, L.; Blank, T. (1988): 'Parallel Logic Simulation on General Purpose Machines,' *25th DAC*, 166–71.

Sparmann, U. (1992): 'Derivation of High Quality Tests for Large Heterogeneous Circuits: Floating Point Options.' *EDAC 92*, 355–60. (Insertion of test functions and cost of testing.)

Stroud, C. E. (1988): 'An Automated BIST Approach for General Sequential Logic Synthesis.' *25th DAC*, 3–8. (System BIST for diagnostic purposes as opposed to for test only. Signatures give diagnostic information.)

Sutlieff, C. (1991): 'Testing Time for ASICs.' *IEE Review*, **37**, 27–31. (Good general introduction to problems and approach, with little technical detail. Good starter.)

Szygenda, S. A.; Thompson, E. W. (1975): 'Digital Logic Simulation in a Time-Based, Table Driven Environment.' *Computer*, **8-3**, March, 24–49 (Two papers.)

Takamatsu, Y.; Kinoshita, K. (1989): 'CONT: A Concurrent Test Generation System.' *Trans IEEE* **CAD-8**, 966–72.

Takamine, Y.; Miyamoto, S.; Nagashima, S.; Miyoshi, M.; Kawube, S. (1988): 'Clock Event Suppression Algorithm of VELVET and its Application to the S-820 Development.' *25th DAC*, 716–19. (Claims suppression of 60% of total events.)

Ulrich, E. (1980): 'Table Look-up Techniques for Fast and Flexible Digital Logic Simulation.' *17th DAC*, 560–63.

Ulrich, E. G.; Baker, T.; Williams, L. R. (1972): 'Fault-Test Analysis Techniques Based on a Logic Simulation.' *9th DAC*, 111–15. (Original paper on concurrent fault simulation.)

Ulrich, E. G.; Baker, T. (1974): 'Concurrent Simulation of Nearly Identical Digital Networks.' *Computer*, **7-4**, Apr., 39–44. (Good explanation – use with Ulrich *et al.* (1972).)

Ulrich, E.; Suetsugu, I. (1986): 'Techniques for Logic and Fault Simulation.' *VLSI Syst. Des.*, **7**, Oct., 68–81. (Excellent review of event driven and fault simulation. Not technical.

Vellandi, B.; Lightner, M. (1992): 'Parallelism Extraction and Program Restructuring of VHDL for Parallel Simulation.' *EDAC 92*, 81–7. (Asserts need to treat event processing as part of parallel system.)

Verhallen, Th. J. W.; van de Goor, A. J. (1992): 'Functional Testing of Modern Microprocessors.' *EDAC 92*, 350–4.

'VHDL Language Reference Manual.' IEEE Standard 1076–1987.

Visweswariah, C.; Chadha, R.; Chen, C.-F. (1988): 'Model Development and Verification for High Level Analogue Blocks.' *25th DAC*, 376–82.

Waicukauski, J. A.; Eichelberger, E. B.; Forlenza, D. O.; Lindbloom, E.; McCarthy, T. (1985): 'Fault Simulation for Structured VLSI.' *VLSI Design*, **6**, Dec., 20–32.

Waicukauski, J. A.; Lindbloom, E.; Rosen, B. K.; Iyengar, V. S. (1987): 'Transition Fault Simulation.' *IEEE Design and Test of Computers*, **4-2**, Apr., 32–8. (Delay fault testing.)

Wallace, D. E.; Sequin, C. H. (1988): 'ATV: An Abstract Timing

Verifier.' *25th DAC*, 154–9.

Wang, L-T.; Hoover, N. E.; Porter, E. H.; Zasio, J. J. (1987): 'SSIM: A Software Levelized Compiled-Code Simulator.' *24th DAC*, 2–8. (Three approaches to LCC with advantages and disadvantages.)

Wang, J-F.; Kuo, T-Y.; Lee, J-Y. (1989): 'A New Approach to Derive Robust Tests for Stuck-Open Faults in CMOS Combinational Logic Circuits.' *26th DAC*, 726–9. (Fairly obvious theorems on avoiding hazards.)

Whetsel, L. (1988): 'A standard Test Bus and Boundary Scan Architecture.' *Texas Instruments Technical Journal*, July/August, 48–59.

Widdoes, L. C.; Harding, W. C. (1984): 'CAE station uses Real Chips to Simulate VLSI-Based Systems.' *Electronic Design*, Mar. 22nd. (Realchip in VALID SCALD System. Good review of reasons for hardware modelling, even when ignoring breadboard section.)

Widdoes, L. C.; Stump, H. (1988): 'Hardware Modelling.' *VLSI Design*, **9**, July, 30–8, 98. (Compares structural, behavioural and hardware modelling.)

Williams, T. W.; Brown, N. C. (1981): 'Defect Level as a Function of Fault Coverage.' *Trans IEEE*, **C-30**, 987–8.

Wong, K. E.; Franklin, M. A.; Chamberlain, R. D.; Shing, B. L. (1986): 'Statistics on Logic Simulation.' *23rd DAC*, 13–19. (Indicates the proportion of time spent in each of the phases of event driven simulation.)

Wunderlich, H-J. (1987): 'On Computing Optimised Input Probabilities for Random Test.' *24th DAC*, 392–8. (Contains some interesting statistics on size of random pattern test sets.)

Yang, H. G.; Holburn, D. M. (1991): 'A Hierarchical Approach to Timing Verification in CMOS VLSI Design.' *EDAC 91*, 266–70.

Index

Bold page numbers indicate term definition or main reference. 'S' implies the start of a Section, 'C' the start of a Chapter.

rising value 169ff
robust test (for delay faults) 241

scan design 44S
scan register test 45f
self test 51S
sensitivity list (VHDL PROCESS) 104
sequential controllability 57
sequential logic testing 42, 44
sequential observability 60f
set-up time 5, 14, 42, 176, 181
set-up time, negative 182
shift register testing 45
signal groups 144
SIGNAL (VHDL) 99, **104**, 161f
signature analysis 51S
silicon compilation 35S
simulation cycle 244f, 254
simulation, dangers in 29, 166, 245
simulation environment 99ff, 158, 171
simulator, desirable features 243
skew 249
slack **204**ff
specification 2, 20, 245, 256
spike (on clock line) 26, 250
spike removal 166, 241, 244, 247
split (group) *see* group
state machine models 158S
state table **130**
strength **172**
strong (strength) 173
structural model **152**, 184ff
stuck at fault model 17, 41
stuck closed 41, 92
stuck open 41, 66, 90S, 249
surrogate fault 211
switch level 11S, 24, 149, 175, 187, 199
synchronous system 113, 120
System Development Engine (SDE, Zycad) 253

table as (model) 151
table based simulation **112**
  see also event driven simulation
technology 243
test bench **98**, 116, 123, 132
test data 96
test reduction 210f
test, scan register 45f
Test Access Port (TAP) 47f
testability 61S, 247
time wheel **126**ff, 144
time wheel overflow 142S
timing verification 139f, 192C, 245
top down design 7, 246
topology (of network) 192
transmission line 34, 250
transport delay **152**, 166, 247, 258
TTL 4, 15, 26, 35, 65, 91, 160, 162, 171, 183, 186
TYPE (VHDL) **98**ff, 171

unknown strength 173
unknown/uncertain value 169f
  see also X

VARIABLE (VHDL) **104**, 161f
vertex (of D-cube) **74**ff

WAIT (VHDL) 98f, 100, 104, **107**
weak (strength) 173
what-if 2, 244, 252
wire gates 34f, 186S
wiring delay 4, 34S, 143S

X value 167, 249

yield 207f
YSE (Yorktown Simulation Engine) 115, 254

Z (high impedance) 171, 187
Zycad 252ff, 256